Textile Technology and Design

Textile Technology and Design

Editor

Abdulkerim Macar

scitus academics

Textile Technology and Design
Edited by **Abdulkerim Macar**

ISBN: 978-1-68117-673-4
Library of Congress Control Number: 2016946539

SCITUS Academics LLC,
616, Corporate Way, Suite 2, 4766,
Valley Cottage, NY 10989

www.scitusacademics.com

Notice

Printed in United States of America on Acid Free Paper

Preface

Textile is one of the promising industries as a regularly used material for ages now and as an industry. There is always the textile industry, as clothing is one of many basic human needs. To make textiles, textile designers often need to design textiles. Textile design is essentially the design process for knitted, woven or possibly printed fabrics. Tissues can be made by weaving and knitting and basic decoration is given by printing. Textile design is a specialized process that includes different methods of textile production. It includes the surface design and the structural design of a textile. The textile designer must have knowledge of the manufacture of yarn, weaving, knitting, dyeing, finishing processes and also of many types of looms, knitting machines and printing. The profound understanding of the technical aspects of production as well as the properties of fibers, yarns and textile colors helps the designer to visualize the end product.

This book outlines the strategies and technologies that have driven the innovation of textile apparel. The textile technology concept is closely linked to both manufacturing and decoration. It addresses the crucial role of the interior at the intersection of design and technology, with a variety of interdisciplinary arguments from a wide range of contributors: from designers to researchers and scholars.

Table of Contents

1

CHAPTER

Modification of Silk Fibres with Acetic Anhydride and Phthalic Anhydride-Effects on Dyeability and Physio-Chemical Properties

Md. Ibrahim H. Mondal [1], Md. Khademul Islam [1], Md. Mofakkarul Islam [1], Md. Obaidul Hoque [1], Md. Asadul Hoque [2], Md. Nojibor Rahman [1]

[1]*Department of Applied Chemistry & Chemical Engineering, Rajshahi University, Rajshahi, Bangladesh*

[2]*Department of Materials Science and Engineering, Rajshahi University, Rajshahi, Bangladesh*

ABSTRACT

Silk fibres were chemically modified with acetic anhydride (AA) and phthalic anhydride (PA) to improve the tenacity, thermal properties, dye-abilities and colourfastness properties. Prior to the modification of silk with anhydrides, optimum modification conditions were established and modification was estimated on the weight gain. Improvements of tenacity were 7.56% for 13.63% acetic anhydride loading and 12.76% for 18.04% phthalic anhydride loading. Thermogravimetric analysis showed the improvement of thermal stability for the acetic anhydride modified silk whereas, phthalic anhydride modified silk showed inferior thermal property compared to that of the unmodified silk. Silk fibres modified with the both anhydrides showed better dye-ability to Direct Red 28 and Direct Orange 31, and improve colour fastness properties to the sunlight.

Keywords: Silk fibre, Anhydride, Dyeability, Tenacity, Thermal properties

1. INTRODUCTION

Silk is continuous filament of protein fibre, for its unique characteristics of strength, fineness, length and luxury appeal, is called the "queen of fabrics" [1-4].

Silk fibres are valued for the outstanding characteristics that have exploited for the production of precious textile goods. However, silk possess some inferior textile performance, such as photo yellowing, an yellowing, wash and wear abrasion resistance, crease proofing properties, wrinkle recovery, rub resistance, colour fastness, dimensional stability, oil and water repellency etc [5, 6].

The characteristics of silk can be improved by chemical modification techniques. Among the various methods available for the modification of silk, chemical modification with anhydride appear attractive and are recently used in textile applications [6]. Modification enhance some properties such as crease recovery, tensile strength and elongation, light resistance, dyeing behavior, colour fastness, wash and wear behavior etc [7, 8].

The physico-chemical properties of modified silk fibres largely depend on the chemical structures and functional side groups of the loaded compound. In this study two structurally different anhydrides such as acetic anhydride and phthalic anhydride were used for modification of degummed silk fibres. A comparative study was made on tensile strength, thermal property, dye-ability and colour fastness properties of modified and unmodified silk fibres.

2. MATERIALS

Silk fibres were collected from Bangladesh Sericulture Research and Training Institute (BSR and TI), Rajshahi, Bangladesh. Reagent grade acetic anhydride (AA, CH_3-CO-O-CO-CH_3), phthalic anhydride (PA, $C_8H_4O_3$), sulphuric acid, sodium carbonate, ammonia solution, sodium hydroxide, sodium sulphate and Direct Red 28 were purchased from BDH (England). Glacial acetic acid, tartaric acid, potassium peroxodisulphate were purchased from Merck (Germany). Phthalic anhydride and Direct Orange 31 were received from Qualikems (India) and Sigma (USA) respectively. All the chemicals were used as received.

3. METHODS

3.1. Degumming of Silk Fibre

Soap solution (wheel soap) of strength 3.5 g/L was prepared and its pH was adjusted to 10.0-10.5 by adding 1 g/L of soda ash (pH change was observed directly with Corning Model-7 pH meter). For each gram of silk fibre 30 mL of liquor was taken and degumming was conducted at a temperature of 95-100°C for 1 hour. During the operation, the pH of the bath falls and then silk fibre was transferred into second bath, and heated for 45 min at the same temperature. After degumming, the silk fibre was washed well with hot distilled water for 3 to 4 times.

3.2. Modification of Silk Fibre

Acid anhydride was dissolved in mixed solvent of water and alcohol, and modifications of silk were performed in that mixed solvent system. The modifying baths were prepared by adding required percentage of an anhydride (90% for acetic anhydride and 80% for phthalic anhydride) and the initiator (10% for acetic anhydride and 8% for phthalic anhydride) on the basis of weight of the fibre. The fibre-liquor ratio was maintained at 1: 30. Modification was started at 30°C and then the temperature was slowly increased almost to the boiling point within about 30 min and continued for 90 min with occasional, then allowed for further 30 min as the bath cools down. During modification, boiling distilled water was added to the modifying baths in order to maintain the fibre-liquor ratio constant. After modification, the fibres were washed with cold distilled water and dried at room temperature [9, 10].

Weight gain or anhydride loading and anhydride efficiency were calculated according to the following formula [9, 10]:

$$\text{Weight gain, \%} = \frac{\text{B-A}}{\text{A}} \times 100$$

$$\text{Anhydride efficiency, \%} = \frac{\text{B-A}}{\text{C}} \times 100$$

where A, B and C are the weight of degummed silk fibre before modification, weight of the silk fibre after modification, and weight of anhydride used, respectively.

3.3. Measurement of Tenacity of Silk Fibres

Tenacity (Breaking strength) of silk fibres was measured by using the tensile strength tester (Torsees Schopper type-OS-100). For measuring the tenacity of silk fibre, at first the denier of the fibre was determined by the following equation.

$$D = \frac{W}{L} \times 9000$$

where D, W and L are the Denier of fibre, weight of fibre in grams and length of fibre in metre, respectively.

Silk fibres were cut into equal pieces with the length of 30 cm and weighed. From the above equation denier was measured for each test sample (fibre) and placed between two jaws of the tensile tester and maintained a length of 10 cm of each specimen. One twist per 2 cm was given along the length of the fibre between the jaws of the machine. The breaking load was observed at the point of break on the scale of the machine. In each experiment, tenacity for 10 specimens was taken and the mean of 10 readings gave the tenacity of the fibre. If 1000 g was required to break down 200 denier fibre then its tenacity will be

$$= \frac{1000}{200} = 5.0 \text{ g/denier}$$

Tenacity of raw, degummed, and modified silk fibres were also measured before and after exposure to sunlight in air.

3.4. Measurement of Thermal Properties

Thermal stabilities of the degummed and modified fibres were determined with the Thermogravimetric Analyser (Seiko-Extar TG/DTA-6300, Seiko, Japan). For the measurement, 4-8 mg of silk fibre was taken in the sample pan and 7.711 mg of alumina was taken in the reference pan. The measurement was conducted upto 600°C with the heating rate of 10°C/min in nitrogen atmosphere.

3.5. Dyeing of Silk Fibres with Direct Dyes

Direct dyestuffs were dissolved by first pasting with cold water and then by adding boiling water. The dye baths were prepared by adding 2% dye (Direct Orange 31 and Direct Red 28), 5% electrolyte and 3 - 5% Na_2CO_3 in aqueous solution on the basis of the weight of silk fibre. The fibre-liquor ratio was maintained at 1 : 30. Before immersing the silk fibre in dye bath it was soaked well in distilled water and squeezed for even absorption of dye particles. Dyeing was started at 40°C and the temperature was then slowly increased almost to the boiling point within about 30 min and continued for 90 min with occasional stirring by a glass rod, then allowed for further 30 min as the bath cools down. During dyeing, boiling distilled water was added to the dye baths in order to maintain the fibre-liquor ratio constant. After dyeing, the fibres were squeezed over dye baths so that not a single drop of exhausted dye liquor was lost, rinsed in cold distilled water and dried at room temperature. The rinsed water was added to the exhausted dye bath and spectroscopic method was used to calculate the dye uptake by the fibres.

3.6. Colour Fastness to Sunlight [12]

Colour fastness test was carried out on raw, degummed, modified and modified dyed silk fibres. Specimens of the fibres were attached separately on a board by a glass rod and placed on the roof of a building for exposure in the open air under the sun without any protection from weathering, but was protected from rain, dews, etc. The specimens were exposed under the sun for seven hours each day continued for 300 h in the month of May to July. After every 50 hours, the fastness was assessed with the Grey Scale by comparing the change in colour of the specimen with that of the standard.

4. RESULTS AND DISCUSSION

4.1. Chemical Modification of Silk Fibre with Acetic Anhydride and Phthalic Anhydride

Degummed silk fibre was chemically modified with acetic anhydride (AA) and phthalic anhydride (PA) and a comparative study was made to see the responses of modified fibres to sunlight in respect of loss of tenacity, colour fastness and thermal properties. As the weight gain of modified fibre considerably dependent on anhydride concentrations, initiator concentrations, temperature and reaction time, modification was conducted at optimized conditions.

It was seen that the percent weight gain of the silk fibre increased considerably with increasing anhydride concentration upto 90% for AA and 80% for PA. Beyond this condition, the percent weight gain started to decrease due to the decreased reactive sites of silk fibres at higher anhydride concentration. The optimized conditions for modification for acetic anhydride were 90% AA, 10% $K_2S_2O_8$, 60 min and 80°C, and those for phthalic anhydride were 80% PA, 8% $K_2S_2O_8$, 80 min and 70°C respectively and are listed in **Table 1**. Under optimized conditions, the maximum weight gains of AA and PA onto degummed silk fibre were 13.79% and 18.16%, respectively.

Table 1. Optimum modification conditions of degummed silk fibres

Anhydrides	Anhydride conc., %	Initiator conc., %	Temp., °C	Time, min	Weight gain, %	Anhydride efficiency, %
Acetic Anhydride	90	10	80	60	13.63	15.32
Phthalic Anhydride	80	8	70	80	18.04	22.70

During modification of silk fibres with anhydrides, the initiator ($K_2S_2O_8$) generates active sites on the backbone of the silk fibres. As the created silk macroradicals are immobile, the weight gain depends on the availability of the anhydride ions or radicals in the vicinity of fibre surface and weight gain become maximum at optimum anhydride and initiator concentrations [13].

Figure 1 shows that percent weight gain and its efficiency of AA and PA onto silk at optimized condition. The maximum weight gain for AA and PA was 13.63% and 18.04% at 80°C and 70°C respectively. At this temperature, the initiator ($K_2S_2O_8$) generates activated anhydrides and silk macroradicals on fibre surface, and combines themselves to give highest conversion. Again, the increase in temperature increased the rate of diffusion of anhydride into the fibre pores. Decreasing of percent weight gain and efficiency of AA and PA beyond the optimum temperature may be the result of degradation of silk during modification [14]. At temperatures over 80°C for AA and 70°C for PA, the reaction became reversible and, accordingly, ceiling temperature is established when the forward and backward reactions became equal. Consequently, at high temperatures, exceeding the optimum values, the weight gain started to decrease.

4.2. Tensile Strength

It is observed from the **Table 2** that tenacity of degummed silk fibre is 29.63 g/denier and modified silk fibre are 31.87 and 33.41 g/denier (for acetic anhydride and phthalic anhydride) respectively. On comparison between degummed and modified silk fibres, it is seen that the tenacity of modified silk fibres are higher than that of degummed silk fibre.

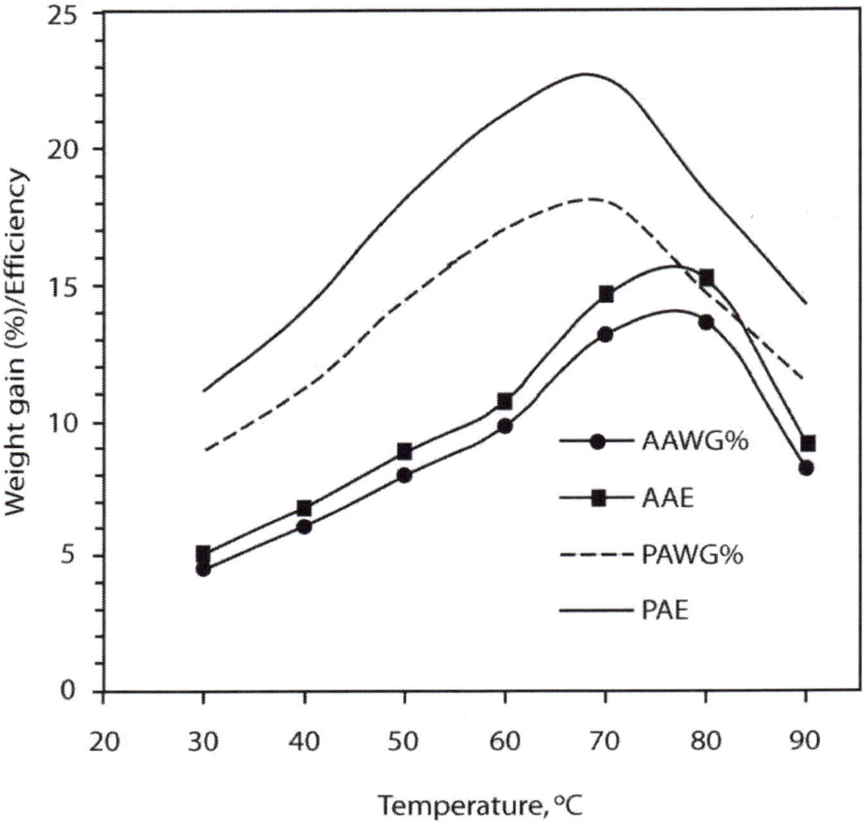

Figure 1. Modification of degummed silk fibres: Bath condition for acetic anhydride (AA) and phthalic anhydride (PA), 90% and 80%; initiator ($K_2S_2O_8$), 10% and 8%, reaction time, 60 and 80 min respectively. WG and E means weight gain percentage and anhydride efficiency

Table 2. Physical and thermal properties of untreated and treated fibres

Silk fibres	Weight gain, %	Tenacity, g/denier	Thermal stability[a]	
			weight loss, %	Temp., °C
Degummed	0	29.63	3-50	282-332
AA modified	13.63	31.87	7-53	292-345
PA modified	18.04	33.41	3–24-58	163-205-347

[a]Determined by TG at the heating rate of 10 °C/min.

This was caused due to weight gain of silk fibre with anhydrides; the crystallinity of silk fibre was reduced due to the incorporation of anhydride with the polypeptide chain and gave an additional strength to silk fibre [15]. Again, the tenacity of phthalic anhydride modified silk fibre is higher than that of acetic anhydride modified silk fibre. It may be explained that higher modification gives the higher strength to silk fibre as well as higher breaking strength of PA modified silk compared to that of AA modified silk fibres.

4.3. Thermal Properties

The thermal behavior of degummed, AA and PA modified fibres were examined by thermogravimetric analysis which is shown in the **Figure 2.** It can be observed from **Figure 2** that the loss in weight for degummed, AA and PA-modified silk fibres are around 3%, 7% and 3 % at 282°C, 292°C and 163.5°C respectively in nitrogen which implies that only moisture were removed with no degradation in these temperature. Actually for degummed silk, T_d^{1st} occurred with 50% weight loss in the temperature range 282-332 °C for AA modified silk, T_d^{1st} occurred with 53% weight loss in the temperature range 292-345°C, whereas, for PA modified silk, first pyrolysis degradation T_d^{1st} occurred in the early stage with 24% weight loss in the temperature range 163-205°C. Second degradation T_d^{2nd} for PA modified silk with 58% weight loss occurred at the temperature range around 300-347°C. For PA modified silk fibre, weight loss was found twice. The first degradation occurred in linkage between silk fibroin and PA unit and, next degradation occurs in fibroin backbone of silk fibre.

From the **Figure 2** it is seen that the thermal stability of AA modified fibre is higher than unmodified fibre; which may be happened due to the incorporation of anhydride molecule onto the silk fibre [16]. The anhydride molecule caused a protective layer onto the active surface of the silk fibre which first worked against the applied heat and ultimately decreased the heating strength on the fibroin backbone. So the modified fibre showed higher thermal stability. **Figure 2** also shows that PA modified silk exhibits the lowest thermal stability among all the modified and unmodified silk fibres. It might be attributed that at the high temperature, phthalic anhydride unit undergoes exothermic degradation which enhances the early degradation of the silk fibroin.

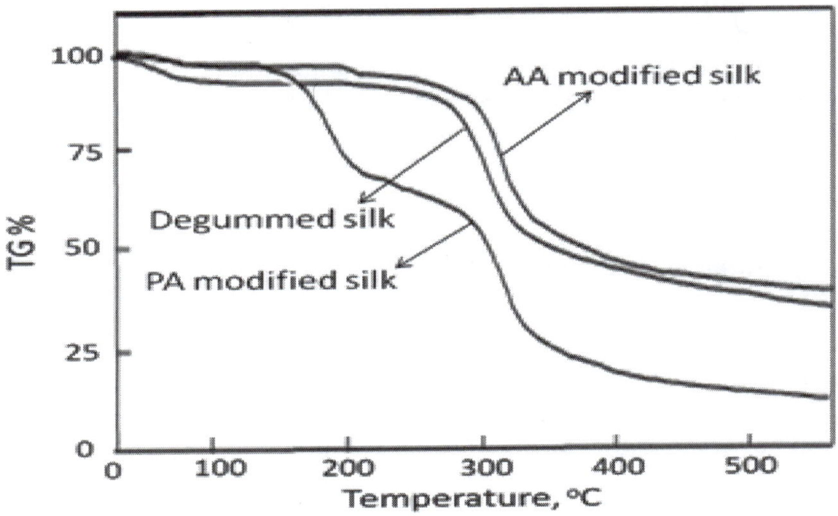

Figure 2. TG analysis of degummed, acetic anhydride and phthalic anhydride modified silk with the heating rate of 10 °C/min

4.4. Dyeing of Modified and Unmodified Silk

The degummed and modified (AA and PA modified) fibres were dyed with Direct Red 28 and Direct Orange 31 with the condition given in the footnote of the **Table 3.** From the **Table 3,** it can be seen that the dye exhaustion of modified silk fibre is higher than that of unmodified (degummed) silk fibre. The anhydride molecule make the fibre hydrophobic, and due to their hydrophobic properties the dye ability of the silk fibres increases. From the experiment it was observed that the dyed modified silk fibre was more lustrous, soft, pleasant at appearance, attractive in hue and gave more uniform shade than the dyed unmodified (degummed) silk fibres in case of all the dyes used.

Table 3. Dyes uptake by the degummed, AA modified and PA modified silk

Name of dye	Dye exhaustion, %		
	Degummed silk fibre	Modified silk fibre	
		Acetic anhydride	Phthalic anhydride
[a]Direct Red 28	70.22	80.91	84.38
[b]Direct Orange 31	77.58	88.83	91.00

Bath condition: Fibre-liquor ratio, 1:30; reaction time, 90 min; conc. of dye, 2%; electrolyte, 5%; and Na_2CO_3 (3%)[b] or (5%)[a] on the basis of the weight of silk fibre.

4.5. Colour Fastness to Sunlight

Raw silk, degummed silk, modified silk and dyed silk fibres were exposed to sunlight for up to 300 h and colour changes were observed in every 50 h exposure. From the **Table 4** it is observed that with the progress of exposure time to sunlight in open air, colour of raw silk fibres changes from golden yellow to dull yellow. But in case of degummed silk fibres, colour changes appears to be very slight from white to yellowish white. That means, raw silk fibre undergoes degradation faster than degummed silk fibre. This is possibly, due to the presence of greater amount sericin in raw silk fibre than that in the degummed silk fibre. Xanthophylls is responsible for yellow colour of raw silk. But the effect of UV-light in air exanthoproteic reaction occurs, hence, yellow colour of raw silk turns to dull yellow. A yellowing mechanism is also postulated that by the effect of UV-light, *p*-hydroxyl- phenylpyruvyl peptide, a chemical component of yellowing of silk is converted to *p*-hydroxyl phenylpyruvic acid and amino acids such as alanine, glycine, valine, leucine etc [17].

From the **Table 4** it is also observed that the colour change of the degummed silk and modified silk (fibroin) is slight compare to raw silk. That was because of the formation of glycidyl radicals in silk fibroin by the irradiation in sunlight, which reacted with atmospheric oxygen and change of peroxy radicals [18, 19]. That oxidation reaction occurred at the earlier time of light exposure. When the reaction was completed the change in colour or fading did not occur. The colour fastness of the modified silk fibre was better than that of degummed silk fibre, because, modification with AA and PA formed a protective layer on the silk surface which itself partially protected UV-radiation and decreased the rate of UV-light penetration into the fibre backbone. This effect decreased the action of the UV-light on the change of the colourfastness. The protective layer of AA and PA modified fibres also reduced the rate of moisture diffusion to the site of fading causing the decreasing rate of yellowing.

Table 4. Colourfastness and change in colour of raw, degummed, AA modified and PA modified silk on exposure to sunlight in air.

Exposure Period, hours	Fastness grade and colour			
	Raw silk	Degummed silk	AA modified silk	PA modified silk
0	5 Golden yellow	5 White	5 Cream white	5 White
50	3 Pale yellow	5 White	5 Cream white	5 White
100	3-2 Pale yellow	5-4 Yellowish white	5 Cream white	5-4 Whiteness decrease
150	3-2 Pale yellow	5-4 Yellowish white	4 Cream white	4 Whiteness decrease
200	2 Dull yellow	5-4 Yellowish white	4-3 Whiteness decrease	4 Whiteness decrease
250	2 Dull yellow	4 Yellowish white	3 Slightly yellowish white	4-3 Slightly yellowish white
300	2-1 Dull yellow	4-3 Yellowish white	3 Slightly yellowish white	3 Slightly yellowish white

Colourfastness properties of degummed and modified silk fibres dyed with Direct Red 28 and Direct Orange 31 are tabulated in the **Table 5**. From the **Table 5**, it is observed that modified dyed silk fibres give better colour fastness properties than that of degummed dyed silk fibre when they are exposed to sunlight in air. It might be explained that after modification, the -NH_2 groups on silk backbone were blocked by modification reaction, as a result the atmospheric oxygen could not react with the –NH_2 group of silk fibre to form peroxy radicals. So, the change in colour or fading did not occur or slightly occurred [20].

Table 5. Colour fastness and change in colour of degummed, AA modified and PA modified silk dyed with Direct Red 28 and Direct Orange 31on exposure

Exposure Period (hrs)	Firstness grade and colour					
	Degummed silk		AA modified silk		PA modified silk	
	Direct Red 28	Direct Orange 31	Direct Red 28	Direct Orange 31	Direct Red 28	Direct Orange 31
00	5 (Bright red)	5 (Bright Orange)	5 (Bright red)	5 (Bright Orange)	5 (Bright red)	5 (Bright Orange)
50	5-4	5-4	5	5-4	5-4	5-4
100	5-4	5-4	5-4	5-4	5-4	5-4
150	5-4	4	5-4	4	4	4
200	4-3	4-3	4	4	4	4
250	3	4-3	4-3	4	4-3	4-3
300	2-1	3	3-2	4-3	3	4-3

5. CONCLUSIONS

Degummed silk fibres were chemically modified with acetic anhydride and phthalic anhydrides with optimum anhydride concentration, initiator concentration, time and temperature. Both the acetic anhydride and phthalic anhydride modified fibres showed superior tenacity, dye-abilities and colour fastness properties compared to that of unmodified silk fibres. Among the modified and unmodified silk fibres, acetic anhydride modified silk showed the best thermal stability and phthalic anhydride modified silk showed the worst thermal stability. Dye-ability of the modified fibres with Direct Red 28 and Direct Orange 31 were higher than unmodified fibres.

The dyed modified fibres were more lustrous, uniform and pleasant at appearance than degummed silk fibre.

REFERENCES

1. Prayag R. S., (1983). Dyeing of Wool, Silk and Man made Fibre, Bombay, 49-77.

2. Joseph, M. L. (1976). Introductory Textile Science (3rd Ed), Holt, Rinbehart and Winston, New York, 13 & 104-111.

3. Sinha, S. (1990). The Development of Indian silk. Oxford and Publishing Co. New Delhi, 1 & 8.

4. Hess, K. P. (1978). Textile Fibres and Their Use (6th Ed), Oxford and IBH Publishing CO., New Delhi, 157-59 & 222-48.

5. Tsukada, M., Goto, Y., Freddi, G., Matsumura, M., Shiozaki, H., Ishikawa, H. (1992). Structure and Physical Properties of Epoxid-treated Tussah Silk Fibres, J. Appl. Polym. Sci., 44, 2203-2211.

6. Tsukada M., Freddi, G., Shiozaki, H., Pusch, N. (1993). Change in Physical Properties of Methacrylonitrile Grafted Silk Fibres, J. Appl. Polym. Sci., 49, 593-598.

7. Corbman, B. P. (1985). Textiles Fibre to Fabric (6th Ed), McGraw Hill Co., New York, 1985, 2, 201, 299 & 301.

8. Deepti, B., Gupta, K. C. (1988). Development of dyeing process of silk with natural dyestuffs, Dept. Cloth Text, G. B. Plant University (India), Colourage, 35(22), 22-24.

9. Rafie, M. H. E., Hafiz, S. A., Hasan S. M., Hebeisha. A. (1994). Grafting of methacrylic acid to loomstate viscose fabric using KMnO4/NaHSO3 system, Ploym. Polymer Compos., 2, 100-1005.

10. Hebeish, A., Katouch A., EL-Ratie, M. H. I. (1971). Graft co-polymerization of vinyl monomers with modified cotton. 11. Grafting of acrylonitile and methyl methacrylate on acetylated cotton, J. Appl. Polym. Sci., 15, 11-14.

11. Giles, H.G. (1974). A Laboratory course in dyeing (3rd Ed), The Society of Dyers and Dolourisits. Bradtord Yordshire, England, 54-58 & 79.

12. Farouqui, F. I., Mondal, Md. I. H., Sheikh, M. R. K (March 1995). Study on the optimum dyeing conditions in dyeing Rajshahi Silk with acid and direct dyes and their colour

fastness properties, Bangladesh chemical society, Abstracts, 56.

13. Mondal, Md. I. H., Hoque, M. A., Latif, M. A., Islam M. K. (2014). Improved characteristics of acetic anhydride and phthalic anhydride modified jute fibers, Res. J. Text. Apperal, 18(1), 64-70.

14. Mondal, Md. I. H.; Haque, Md. M.-U. (2007). Effect of grafting with methacrylate monomers onto jute constituents using potassium persulfate initiator catalysed by Fe(II), J. Appl. Polym. Sci, 103, 2369-2375.

15. Haque, Md. M.-U., Zaman, M. A. A., Rahaman, M. H., Hossain, M. Z., Maniruzzaman, M. (2014). Thermal and tensile mechanical behavior of acetic anhydride treated silk fibres, Intern. J. Mats. Sci. Applications, 3(3), 106-110.

16. Ishikawa & Co-Workers (1969). Differential Thermal Analysis of Silk fibre, Sen-i Gakkaishi, 25(9), 425-31.

17. Trotman, R. (1984). Dyieng and Chemical Technology of Textile Fibres, 6th ed, EBLS London, 447-456.

18. Li, G., Liu, H., Li, T.D., Wang, J. (2012). Surface modification and functionalization of silk fibroin fibers/fabric toward high performance applications, Mats. Sci. Eng., C32, 627–636.

19. Tsukada, M., Goto, Y., Freddi, G., Shiozaki, H. (1992). Chemical modification of silk with aromatic acid anhydrides, J. Appl. Polym. Sci., 45(7), 1189–1194.

20. Uddin, Md. K., Hossain, M. S. A. (2010). Comparative Study on Silk Dyeing with Acid Dye and Reactive Dye, Intern. J. Eng. Technol., 10(6), 22-27, 2010.

2

CHAPTER

A Smart Multifunctional Water-Oil Repellent Finishing and Dyeing of Cotton / Polyamide 6 Blended Fabrics

E. R. El-Zairy, W. R. El-Zairy

Faculty of Applied Arts, Printing, Dyeing and Finishing Dept. Helwan Univ., Cairo, Egypt

ABSTRACT

Polyamide/cellulosic blends offer valuable possibilities for combination desirable performance properties, i.e. self-cleaning, water-oil repellent, crease recovery angle, handle, etc. The main aim of this study is to search for a novel practical method of carrying out water-oil repellent finishing/reactive dyeing of cotton/polyamide-6 (50-50) in one stage using knittex® FEL crosslinking agent, OLEPHOBOL®CP-U, water/oil repellent agent, trichloroacetic acid as a catalyst as well as reactive dyes of Remazol type. The treated fabrics were padded and then squeezed to a wet pick-up of 70%. After being dried, the fabrics were thermofixed at different temperatures (100-180°C) for various durations (1-5). Good results of the overall fastness properties were obtained ranging between 4 to 4-5.

Keywords: Polyamide/cotton, Dyeing, Finishing

1. INTRODUCTION

Demands for comfort and healthy lifestyles make it desirable to have textile products with multifuncitonality. Incorporated function properties to textiles such as durable press, self-cleaning, antimicrobial and thermochromic characteristics, water and oil-repellent, all have potential industrial applications. [1-6]

Blends of polyamide man-made fibres, with Cellulosic fibres offer valuable possibilities for combining desirable performance properties, developing attractive appearance, and novel design. Recently functional textiles have attracted much attention in recent years, e.g. UV-protection, antimicrobial, anti-radiation, water-and oil repellent, self-cleaning, etc. [7-12] Although the importance of these blends is understood, they have not maintained their reasonable market share. This can be attributed in part to difficulties in wet processing including dyeing to union shades the prevailing problem assiated with the union dyeing of polyamide-6 and cellulose blends from the same dyebath is the preferential uptake dye by polyamide-6. Cellulose/polyamide blends are usually dyed by the two-bath method: for example in a first step by the pad-steam process to dye the cotton component and then by the exhaust process to dye the polyamide component, or vice versa. However, this process is time-consuming and laborious when dyeing large yardages. Single bath method are available whereby two diffferent reactive dyes (one substantative to the polyamide and the other substantative to the cellulose-) can be used to dye this blended fabrics but the result is that the cellulosic component is dyed perfectly satisfactorily but the polyamide component is only superficially dyed, in fact merely stained, i.e. the dye has not diffused into the fibres under this method. [13, 14] Trichloroacetic acid decomposes at about 100°C to give a free radical, in the following manner:

$$C\,Cl_3\,CO\bar{O}\;\overset{+}{H} \longrightarrow {}^{\bullet}C\,Cl_3 + CO_2 + \overset{+}{H} + e$$

$$\overset{+}{H} + {}^{\bullet}C\,Cl_3 \longrightarrow CHCl_3$$

Since the chloroform and carbon dioxide produced are volatile, they are rapidly removed from the dye bath. The striking influence of the chlorine atom(s) on the dissociation of the proton is simply due to its inductive effect, since the chlorine atom in C-Cl attracts electrons in comparison with hydrogen atom in C-H, the effect is relyed in the following manner:

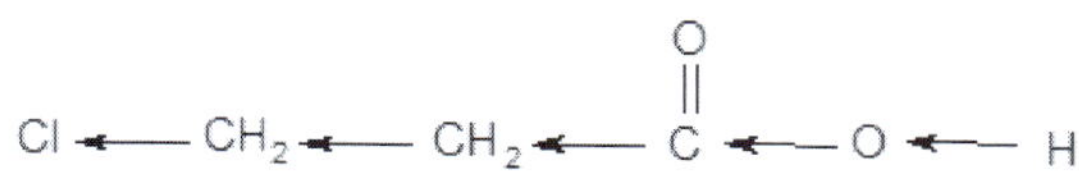

I

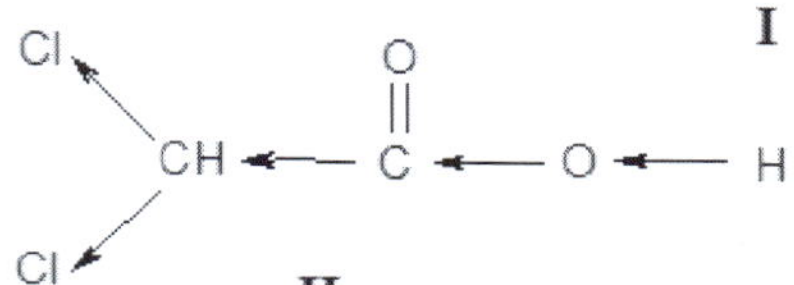

II

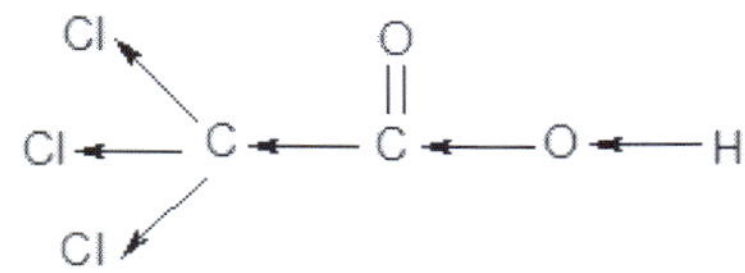

III

The decomposition of trichloroacetic acid, in either boiling dyebaths or during steaming and thermofixation giving a change in pH conditions from acidic to alkaline has been mentioned by Fabwerke Hoechst. [20-21]

In this work, we put forward a new method of carrying out water/oil repellent finishing and reactive dyeing of polyamide-6/cotton blend (Ny/Co) (50/50) in one step to evaluate the impact of the combined treatments on the functional and dyeing properties of the blended fabrics using the ability of trichloroacetic acid (TCAA) to change of the pH of the medium as the temperature changed, as well as to provide a process for dyeing cellulose/polyamide blends, which process makes it possible to dye both types of fibre continuously in a single bath and in one operation of dyeing and finishing.

2. EXPERIMENTAL

2.1. Materials

Mill-scoured, and semi-bleached cotton/polyamide-6 blended fabric (50/50 in yarn, 120 gm²) was used.

Knittex® FEL cross linking agent (reactant crosslinking agent based on a modified dimethylol dihydroxyethylene urea, Huntsman), and oil, water and stain repellent agent, based on dispersion of a polmeric perfluorinated compound, nonionic/cationic, Huntoman, were of technical grade.

Commercial reactive dyes namely: Remazol Red X3BS, Remazol Blue RC, Remazol Golden Yellow RC as well as Remazol Blue RC, Dystar, were used.

Trichloroacetic acid (TCAA) and Na-carbonate were of laboratory reagent grade.

Hostapal® CV-ET (nonionic wetting agent based on alkylarylpolyglycolether, clariant) was of commercial grade.

2.2. Methods

2.2.1. Concurrent Functional of Ny/Co Fabrics

Bifunctional finishing and dyeing of polyamide-6/cotton (Ny/Co) blends were carried out as follows: the blended fabrics samples were padded twice through solution containing Remazol X3BS (20g/l), knittex® FEL crosslinking agent (0-80 g/l). OLEOPHOBOL® CP-U, oil, water and stain repellent agent (0-40 g/l) and trichloroacetic acid (TCAA) (0-4 g/l), the fabrics were then squeezed to a wet pick up of 70%. After being dried at 100°/3min., the fabrics (Ny/Co) were thermofixed at varying temperatures (100-180°C) and for different durations (1-5 min.). Finally, the fabric was washed with cold, hot water then soaped with a solution containing nonionic wetting agent (Hostapal CV-ET) and 2g/l Sodium Carbonate followed by washing with hot, cold water and finally dried.

2.2.2. Testing and Analysis

2.2.2.1. Colour Strength

The colour strength K/S, of the dyed/ finished fabric samples were measured at the wave length of maximum absorbance using an automatic filter spectrophotometer and calculated by the Kubelka – Munk equation [24]:

$$K/S = \frac{(1-R)^2}{2R}$$

Where K is the absorption coefficient, S is the scattering coefficient, and R is reflectance of dyed/finished fabric samples at wavelength of maximum absorption, the highest the K/S value, the greater the dye receptivity of the fabric.

2.2.2.2. Fastness Properties

Evaluation of washing, rubbing, perspiration and light fastness properties were performed according to AATCC test methods: (61-1972), (8-1972), (15-1973) and (16A-1972) respectively.

2.2.2.3. Crease Recovery Angle

Crease recovery tester, type: FF.07. metrimpex, Budapest was used for the determination of the crease recovery angle.

2.2.2.4. Tensile Strength and Elongation

Tensile strength and elongation (warp and weft) was measured on the tensile strength testing machine type FMGW 500 (Veb. Thuringer industriewerk Raunstein) at 25°C and 66% relative humidity.

2.2.2.5. Repellency Rating

Water/oil repellency was evaluated according to AATCC test method 22-1989.

3. RESULTS AND DISCUSSION

Since the primary objective of this work was to investigate the technical feasibility of carrying out dyeing and multi-functional finishing of polyamide-6/cotton fabric blend in a single-step process. The variable studied include trichloroacetic acid (TCAA) as a catalyst, finishing agent, repellent agent as well as thermo fixation time and temperature to discover the optimal condition for currying out bifunctional finishing and dyeing simultaneously of polyamide-6/cotton blended fabric using different coloures of remazol dyes, results obtained along with their appropriate discussions are given below.

3.1. Trichloroacetic Acid Concentration (TCAA)

It is established that reaction of finishing agents with polyamide-6/cotton blended fabrics as well as the reactive dye must be catalysed [25]. Thus the efficiency of simultaneous dyeing and finishing would be expected to rely on the nature and amount of the catalyst used. Table 1 shows the effect of trichloroacetic acid (TCAA) concentration on the extent of simulataneously functionalization and dyeing of polyamide-6/cotton blended fabrics. Obviously, within the range examined, it is clear that, increasing trichloroacetic acid (TCAA) concentration up to 2g/l brings about an improvement in K/S values, crease recovery angle (CRA), handle, water oil repellency rating and overall fastness properties. In addition, increasing the trichloroacetic acid concentration beyond 2g/l results in a slight dcreament in the aforementioned properties, which be discussed in terms of highest penetration of the dye onto polyamide-6/cotton blend along degree of temperature during drying and curing (100° - 160°C) is reached, the acid disappears from the system by free radical decomposition reaction. Further increase in trichloroacetic acid (TCAA), i.e. beyond 2g/l has practically a marginal decreasing effect on the K/S values, and other performance properties, which may attributed to the hydrolysis of reactive dye. Therefore 2g/l TCAA was chosen as the optimal concentration of TCAA in bifunctional finishing and dyeing of Ny/Co blends. Since attachment of the reactive dye to cotton component in the blend occurs through reaction of the dye either with N-methylol finishing agent, which is already chemically bonded to cotton in acidic medium or the ractive dye attachment to polyamide-6/cotton blends in alkaline medium (Equi.1, 2, 6, 7, 8). During the heat treatment of curing, decomposition via demethylolation of N-methylol compound occurs, depending on the catalyst concentration. As a result, amide groups are formed [26]. That is enhancement in colour strength (K/S) by increasing the catalyst concentration could be associated with (a) reactive of the reactive dye with the methylol groups of the N-methylol finishing agent which required high catalyst concentration and (b) decomposition of N-methylol compound at high catalyst concentration which leads to amide groups with high reactivity of the reactive dye (Eq. 3,4,5,6,9).

Table 1. Effect of Trichloroacetic acid (TCAA) concentration on the functional dyeings of polyamide 6/ cotton blended fabrics

Trichloro acetic acid (TCAAg/l)	K/S	Washing at 95°			Rubbing		Perspiration						Light fastness	Handle	Grease Recovery Angle				Tensile Strength		Elongation %		Repellency Rating	
							Acidic			Alkaline					Dry		Wet							
		Alt.	C	N	Dry	Alt.	Alt.	C	N	Alt.	C	N			W.D	F.D.	W.D	F.D.	W.D	F.D.	W.D	F.D.	Water	Oil
0	4.57	2	2-3	2	3	2-3	3	3	2-3	3	2-3	3	5	S	60	78	100	118	40	30	25	35	0	0
1	10.20	2-3	3-4	3	3-4	3	3-4	3-4	3	3-4	3	3-4	6	S	120	141	125	146	39	29	24	34	50	3
2	14.40	4	4	4	4	3-4	4	4-5	3-4	4	3-4	4	6-7	S	130	151	135	152	37	29	23	33	70	5
3	14.00	3-4	4	4	4	3-4	4	4	3-4	3-4	3-4	3-4	6-7	S	130	148	133	149	36	28	23	30	70	5
4	9.22	3-4	3	3-4	3-4	3	3-4	3	3-4	3-4	3	3-4	6	H	131	148	131	148	35	27	21	29	70	5

K/S: Colour strength Alt.: Alteration C: Staining on cotton N: Staining on polyamide-6
W.D.: Warp direction F.D: Weft direction S: Soft H: Harsh
Padding Bath Contents: Knittex® FEL (50g/l), Remazol RedX3BS (20g/l), Water and Oil Repellent (20g/l). TCAA (0-4 g/l); Wet Pick up (70%), Drying at 100°C/ 3min, Curing at 170°C/ 2 min.

Water/oil repellent, tactile and anticrease function as well as the dyeing properties of the simultaneous dyed/finished fabric samples are given in Table 1, the obtained data demonstrate that: (I) using of high molecular weight film-forming polymers, i.e. water and oil repellent agent in conjunction with a fibre crosslinking resin for the production of durable press fabric has been the target of enhancement crease recovery angle, water/oil repellent, handle and overall fastness properties. II) incorporation of the used remazol dyes in finishing formulation is accompanied by a significant improvement in colour yield (K/S) of the obtained dyeings along with a subsequent enhancement in their functional properties which due to the inhibition effect of pH value of trichloroacetic acid (TCAA) during heat treatment as well as inactivation effect of the used Remazol dyes.

$$\text{Cell.OH} + \text{Knittex FEL} \xrightarrow[\Delta]{H^+} \text{Cross linked cellulose} \qquad (1)$$

$$\text{HX - Ny - XH} + \text{Knittex FEL} \xrightarrow[\Delta]{H^+} \text{Crosslinked polyamide 6(Ny)}$$

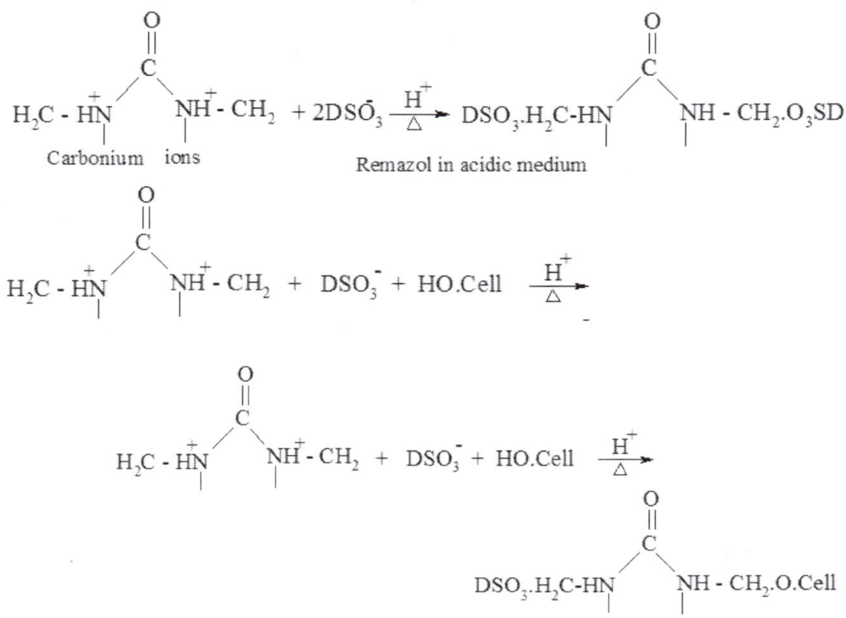

(4)

$$D.SO_2.CH_2.CH_2.OSO_3Na \xrightarrow[\Delta]{O\overset{+}{H}} D.SO_2.CH = CH_2 + NaHSO_4$$

Reactive dye Vinyl form

$$D.SO_2.CH_2.CH_2.OSO_3Na \xrightarrow[\Delta]{O\overset{+}{H}} D.SO_2.CH_2.CH_2.OS\ \bar{O}_3\ Na^{+}$$

Reactive dye Ionized form

(6)

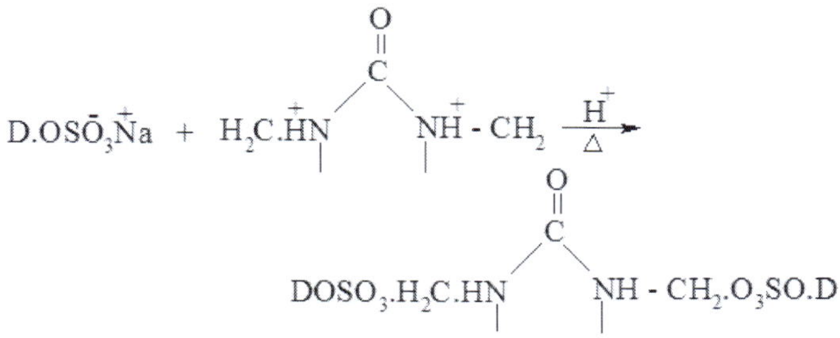

(7)

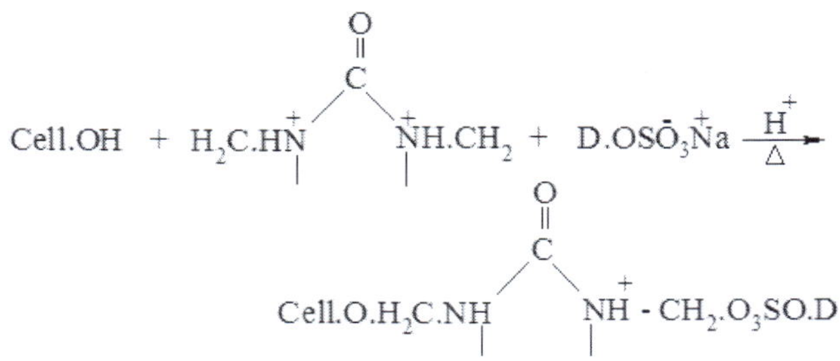

(8)

Where XH = -NH2, = NH-

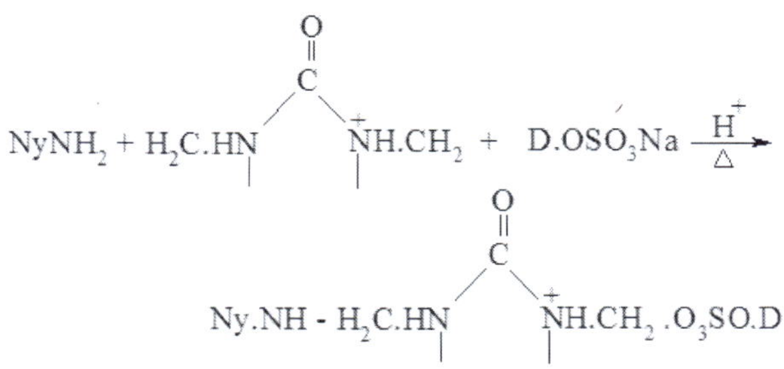

(9)

$$\underset{\substack{\text{polyamide 6}\\ \text{(Ny)}}}{\text{Ny NH}_2} + \underset{\text{Vinyl sulphone}}{\text{D.SO}_2\text{.CH}_2\text{.CH}_2\text{.OSO}_3\text{Na}} \xrightarrow[\Delta]{\overline{\text{OH}}} \text{D.SO}_2\text{.CH}_2\text{.CH}_2\text{.NH.Ny} \quad (10)$$

$$\text{Cell.OH} + \text{D.SO}_2\text{.CH}_2\text{.CH}_2\text{.OSO}_3\text{Na} \xrightarrow[\Delta]{\overline{\text{OH}}} \text{D.SO}_2\text{.CH}_2\text{.CH}_2\text{.O.Cell} \quad (11)$$

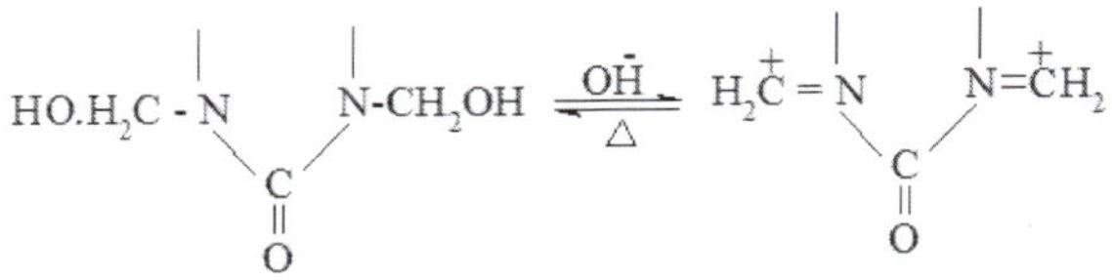

(12)

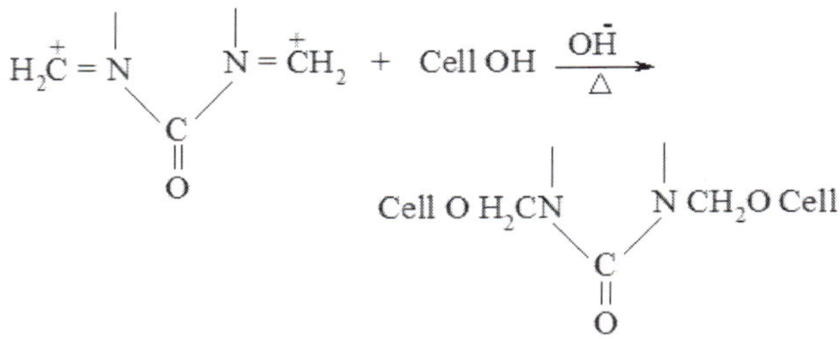

(13)

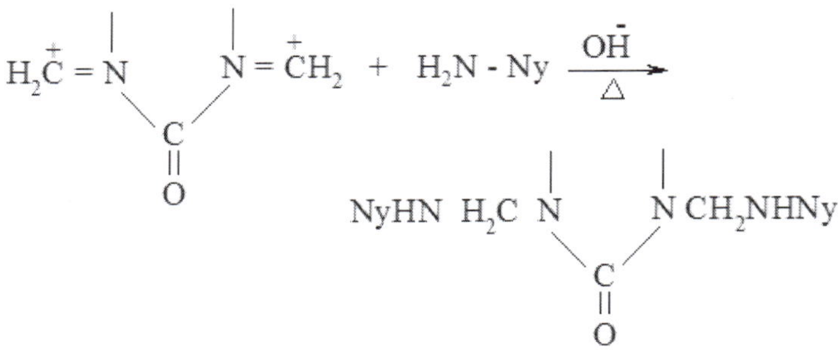

(14)

3.2. Finishing Agent Concentration

Table 2 shows the effect of increasing Knittex® FEL (0-80 g/l) as a finishing agent on the extent of crosslinking. It is clear that: (I) increasing Knittex® FEL concentration up to (60g/l) in the presence of water and oil repellent agent (20g/l), Remazol Red X3BS (20g/l), and trichloroacetic acid (2g/l) as a catalyst is accompanied by an improvement in both the crease recovery angle (CRA) and repellency rating as well as the extent of the multifunctional finishing and dyeing regardless of the used blend. (II) the extent of crosslinking as well as bifunctional properties are determined by the nature and reactivity of the substrate, it is clear from table 2 that, the Knittex ® FEL concentration increases up to 60g/l K/S values of the obtained blended fabric samples tend to increase gradually, which can be attributed to the introduction of new reactions, i.e. reaction of N-methylol finishing agent and reactive dye with polyamide 6/cotton components either in acidic or in alkaline medium due to decomposition of trichloro acetic acid (TCAA) during heat treatment, thereby improving and enhancing the extent of union reactive dye of both the polyamide 6 and Cellulosic components. A further increase in finishing agent concentration i.e., beyond 60g/l has a negative effect that can be discussed in terms of higher extent of side interaction and hydrolysis of reactive dye, thereby minimizing the availability of reactions onto finishing and dyeing/fabric matrix, irrespective of substrate used. It is also clear from (Table 2) that increasing the finishing agent concentration up to 60g/l results an increase in the extent of crosslinking expressed the crease recovery angle values, followed by a slight decrease in the aforementioned properties at higher concentration, i.e., beyond 60g/l, which could be attributed to (1) insufficient diffusion and penetration of the reactant resin into the fabric structure, (2) formation of single ended reactions rather than inter crosslinking due to variation of pH values during heat treatment.The obtained data also demonstrate that: (1) incorporation of the used reactive dyes in the finishing formulation is accompanied by a significant improvement in the colour yield K/S, overall fastness of the obtained dyeings along with a subsequent enhancement in their functional properties namely water and oil repellenty as well as easy care properties, regardless of the used dye (II) overall fastness properties are at the range of 3-4 to 4 using Knittex®FEL concentration up to 60g/l, (III) the enhancement in the functional finishing properties and overall fastness properties of the obtained dyeings most probably due to the increase in the carbonium immonium ion concentration in the proximity of the immobile active sides e.g. OH, NH_2, NH etc., in the polyamide 6/cotton structure thereby giving rise to higher extent of crosslinking [27, 28].

Table 2. Effect of Knittex® FEL on bifunctional finishing and dyeing of polyamide 6/cotton (Ny/Co) blended fabrics using remazol red X3BS as a single dye

Knittex® FEL concentration g/l	K/S	Washing at 95°			Rubbing		Perspiration						Light fastness	Handle	Crease Recovery Angle				Tensile Strength		Elongation %		Repellency Rating	
							Acidic			Alkaline					Dry		Wet							
		Alt.	C	N	Dry	Alt.	Alt.	C	N	Alt.	C	N			W.D	F.D.	W.D	F.D.	W.D	F.D.	W.D	F.D.	Water	Oil
0	5.80	2-3	2	2-3	2-3	2	2-3	3	2-3	3	2-3	2-3	5	S	63	71	95	115	41	32	27	37	50	3
20	8.90	2-3	2-3	3	3	2-3	3-4	3-4	3-4	3-4	3	3	6	S	122	130	128	150	38	31	26	36	70	5
40	14.30	3	3	3-4	3-4	3	3-4	4	4	3-4	3-4	3-4	6-7	S	129	135	135	155	37	30	25	35	70	5
60	16.80	4	4	4	4	3-4	4	4	3-4	4	3-4	4	6-7	S	133	144	140	162	36	30	24	33	80	7
80	16.00	3-4	3-4	3-4	3-4	3	4	3-4	3-4	4	3	3-4	6-7	H	132	143	138	160	36	28	22	32	80	7

K/S: Colour strength   **Alt.**: Alteration   C: Staining on cotton   N: Staining on polyamide-6
W.D.: Warp direction   F.D: Weft direction   S: Soft   H: Harsh
Padding Bath Contents: Knittex® FEL (10-80g/l), Remazol RedX3BS (20g/l), W/O Repellent (20g/l). TCAA (2g/l); Wet Pick up (70%), Drying at 100°C/ 3min, Curing at 170°C min.

Thus it can be seen that using 60g/l Knittex® FEL was sufficient to obtain a Ny/Co fabric with favorable functional and performance properties.

3.3. Water and Oil Repellent Concentration

From table 3 it is clear that: within the range examined, increasing the concentration of repellent agent from zero up 30g/l is accompanied by a significant enhancement in colour strength, overall fastness properties handle, wrinkle recovery and repellent rating, beyond 30g/l concentration no improvement in the aforementioned properties, this is due to I) a direct consequence of the increased the extent of the formation of a hydrophobic/olephobic polymer film on the dyed fabric surface thereby lowering the fabric surface energy and hence repellent properties [29, 30], II) incorporation of the nominated W/O repellent in the padding bath has a positive impact on the depth of obtained dyeings along with a significant enhancement in the fastness properties ratings as a direct consequence of increasing the extent of entrapment and entanglement of the dye molecules within the crosslinked polymer film onto the fabric surface, III) W/O repellency ratings remain unaltered upon increasing the concentration from 30 to 40 g/l, V) TS and elongation decrease upon increasing the concentration of repellent agent from zero up 40g/l, due to increasing the extent of fabric cross-linking and embitterment. Further increase in the W/O repellent up to 30g/l gave soft handle, and harsh handle obtained beyond 30g/l as well as decreasing in T.S. and elongation. This is certainly attributed to the deposition and formation of a repellent film on the fibre surface thereby reducing the energy at fiber surfaces, and reducing surface inter-fibers and inter-yarn frictions.

Table 3. Effect of water/oil repellent concentration on bath dyeing and functional properties of polyamide 6/cotton blends

W/O repellent agent (g/l)	K/S	Washing at 95°			Rubbing		Perspiration						Light fastness	Handle	Crease Recovery Angle				Tensile Strength		Elongation %		Repellency Rating	
							Acidic			Alkaline					Dry		Wet							
		Alt.	C	N	Dry	Alt.	Alt.	C	N	Alt.	C	N			W.D	F.D.	W.D	F.D.	W.D	F.D.	W.D	F.D.	Water	Oil
0	12.90	3	3	2-3	3	2-3	3	2-3	2-3	3	2-3	2-3	5	S	64	73	105	131	40	32	28	38	0	0
10	15.80	3-4	3-4	3	3-4	3	3-4	3	3	3-4	3	3	6	S	124	134	130	155	39	31	27	37	50	3
20	16.20	4	4	3-4	3-4	3-4	4	3-4	3-4	4	3-4	3-4	6-7	S	130	147	138	160	38	31	26	36	70	5
30	17.70	4-5	4-5	4	4	4	4	4	3-4	4	3-4	4	6-7	S	139	158	143	167	38	30	25	35	80	7
40	17.60	4-5	4-5	4-5	4-5	4	4-5	4	4	4-5	4	4	6-7	H	138	157	140	165	36	29	25	34	80	7

K/S: Colour strength   Alt.: Alteration   C: Staining on cotton   N: Staining on polyamide-6
W.D.: Warp direction   F.D: Weft direction   S: Soft   H: Harsh
Padding Bath Contents: Knittex® FEL (60g/l), Remazol RedX3BS (20g/l), Water/Oil Repellent (0-40 g/l). TCAA (2g/l); Wet Pick up (70%), Drying at 100°C/ 3min, Curing at 170°C/2 min.

Table 4. Effect of curing temperature on dyeing and functional properties of polyamide 6/cotton blended fabrics

Curing Temp. °C	K/S	Washing at 95°			Rubbing		Perspiration						Light fastness	Handle	Crease Recovery Angle				Tensile Strength		Elongation %		Repellency Rating	
							Acidic			Alkaline					Dry		Wet							
		Alt.	C	N	Dry	Alt.	Alt.	C	N	Alt.	C	N			W.D	F.D.	W.D	F.D.	W.D	F.D.	W.D	F.D.	Water	Oil
100°C	14.50	3-4	3	3	3-4	3	3-4	3	3	3-4	3	3-4	5	S	90	98	114	137	39	34	27	37	50	3
120°C	14.20	4	3-4	3-4	4	3-4	4	3-4	3-4	4	3-4	4	6	S	130	149	136	165	38	33	25	37	70	5
140°C	16.30	4-5	4	4	4-5	4	4-5	4	4	4	4	4-5	6-7	S	139	160	140	170	37	32	24	36	80	7
160°C	19.60	4-5	4-5	4	4-5	4	4-5	4-5	4-5	4-5	4	4-5	7	S	143	167	148	175	36	32	24	35	80	7
180°C	19.80	4-5	4-5	4-5	4-5	4	4-5	4-5	4-5	4-5	4	4-5	7	H	142	165	146	173	35	30	23	34	80	7

K/S: Colour strength Alt.: Alteration C: Staining on cotton N: Staining on polyamide-6
W.D.: Warp direction F.D: Weft direction S: Soft H: Harsh
Padding Bath Contents: Knittex® FEL (60g/l), Remazol RedX3BS (20g/l), Water/Oil Repellent (30 g/l), TCAA (2g/l); Wet Pick up (70%), Drying at 100°C/ 3min, Curing at 180°C/2 min.

3.4. Curing Temperature

The curing temperature is regarded as one of the most important factors in simultaneous multi-functional finishing and dyeing. Bearing this in mind, it was felt that the curing of the multi-finishing and dyeing samples must becarried out over a range of temperature to discover the optimal temperature, viz, 100°, 120°, 140°, 160° and 180°C for 2 minutes.

Crease recovery angle (CRA) and W/O repellent function as well as the dyeing properties of the simultaneously dyed/finished fabric samples are given in Table 4, the obtained data demonstrate that: I) Incorporation of the used Remazol dyes in finishing formulation is accompanied by a significant improvement in the colour yield, K/S, performance properties of the obtained dyeings along with a subsequent enhancement in their functional properties namely W/O repellent expressed by (50-80), (3-7) as well as CRA expressed by dry warp direction (90-142), dry weft direction (98-165) and wet warp direction (114-146), wet weft direction (137-173), regardless of the used dye II) fastness to washing, rubbing and perspiration of the dyeings are in the range 3 to 4-5, III) handling of the obtained dyeings through increasing the curing temperature from 100 to 160°C gave soft handle, beyond 160°C gave harsh handle.(V) Increasing the temperature from 100° to 160°C, that is the favorable effect of temperature on cross-linking cellulose chains with DMDHEU (Knittex FEL) as well as the progressive consumption of some DMDHEU in binding OLEPHOBOL®CP-U onto cotton structure through its labile hydrogen of urethane groups [31], for example, as illustrated in the following equation:

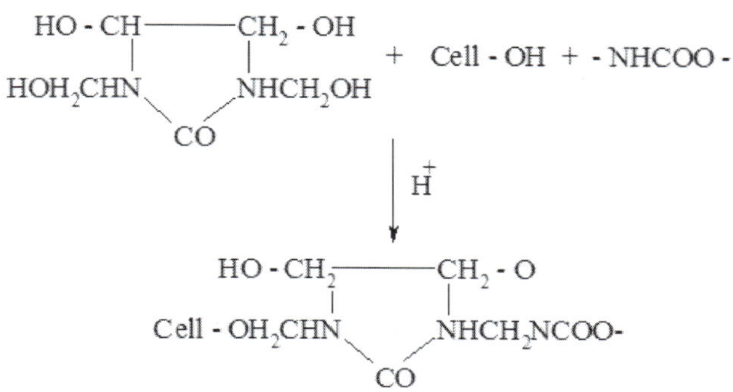

Further raising in curing temperature, i.e. beyond 160°C, has practically no or marginal effect on the aforementioned properties, which may attributed to the enhancement of side interaction and/or partial hydrolysis of substrate/ finishing agent bonds [32].

3.5. Curing Time

To avoid under-cure and over-cure, it becomes inevitable to investigate the effect of curing time on multifunctional finishing and performance properties. Curing the treated specimens were conducted at 160°C for different periods of time ranging from 1 to 5min for achieving a solid shade on polyamide-6/cotton blended samples (Table 5). An enhancement in functional finishing properties, colour strength (K/S) as well as performance properties by increasing duration of curing is somewhat expected. Reasons for this are the following a) cellulose hydroxyls of the cotton component in the blend is immobile. Their reaction with N-methylol finishing agent/ dye system would depend upon the time allowed for the molecules of this system to diffuse inside the fabrics, b) W/O repellent ratings, CRA, K/S, and performance properties attain maximum values within the time used i.e., (3 min), and prolonging it up to 5 min has no effect on these values, which suggests that most of the W/O repellent agent is located at the fabric surface with little penetration inside the fabric structure [33], c) CRA increases upon increasing time to 3 min then remain nearly constant or slightly decreases by raising the time up to 5 min., d) TS and elongation decrease by raising the time from 1 to 5 minutes, due to increasing the extent of fabric cross-linking and embitterment, e) curing at 160°C for 4 and 5 min have practically slight or no effect on the aforementioned parameteries. Obviously curing for 3 min at 160°C appears appropriate for attaining better performance properties.

Table 5. Effect of curing time on dyeing and finishing properties of polyamide 6/cotton blended fabrics

Curing Time (min)	K/S	Washing at 95°			Rubbing		Perspiration						Light fastness	Handle	Crease Recovery Angle				Tensile Strength		Elongation %		Repellency Rating	
							Acidic			Alkaline					Dry		Wet							
		Alt.	C	N	Dry	Alt.	Alt.	C	N	Alt.	C	N			W.D	F.D.	W.D	F.D.	W.D	F.D.	W.D	F.D.	Water	Oil
1	14.60	4	3-4	4	3-4	3-4	3-4	3-4	4	3-4	3-4	3-4	6	S	120	141	123	143	39	33	27	37	50	3
2	18.95	4.5	4-5	4	4-5	4	4-5	4-5	4-5	4-5	4	4	4-7	S	140	168	142	172	37	32	26	36	70	5
3	20.60	4.5	4-5	4-5	4-5	4-5	4-5	4-5	4-5	4-5	4-5	4-5	7	S	149	173	152	178	36	31	24	34	80	7
4	20.70	4.5	4	4-5	4-5	4-5	4-5	4-5	4-5	4-5	4-5	4-5	7	S	147	172	151	176	36	31	24	34	80	7
5	20.75	4.5	4	4-5	4-5	4-5	4-5	4-5	4-5	4-5	4-5	4-5	7	H	145	169	150	172	35	30	23	33	80	7

K/S: Colour strength Alt.: Alteration C: Staining on cotton N: Staining on polyamide-6
W.D.: Warp direction F.D: Weft direction S: Soft H: Harsh
Padding Bath Contents: Knittex® FEL (60g/l), Remazol RedX3BS (20g/l), Water/Oil Repellent (30 g/l). TCAA (2g/l); Wet Pick up (70%), Drying at 100°C/ 3min, Curing at 160°C/ at different time (1-5 min).

Table 6. Technological evaluation of bifunctional finishing and dyeing properties of polyamide 6/cotton blends using different remazol dyestuffs

Dyestuff	K/S	Washing at 95°			Rubbing		Perspiration						Light fastness	Handle	Crease Recovery Angle				Tensile Strength		Elongation %		Repellency Rating	
							Acidic			Alkaline					Dry		Wet							
		Alt.	C	N	Dry	Alt.	Alt.	C	N	Alt.	C	N			W.D	F.D.	W.D	F.D.	W.D	F.D.	W.D	F.D.	Water	Oil
Remazol Red RC	19.75	4-5	4	4	4	3-4	4-5	4-5	4	4-5	4	4-5	6-7	S	147	175	150	179	39	31	30	39	70	5
Remazol Blue RC	20.55	4-5	4	4	4-5	4	4-5	4-5	4	4-5	4	4-5	6-7	S	145	173	149	178	41	30	31	43	70	5
Remazol Golden Yellow RC	13.50	4-5	4-5	4	4-5	4	4-5	4-5	4	4-5	4	4-5	7	S	148	177	149	179	42	31	31	44	80	7

K/S: Colour strength Alt.: Alteration C: Staining on cotton N: Staining on polyamide-6
W.D.: Warp direction F.D: Weft direction S: Soft H: Harsh
Padding Bath Contents: Knittex® FEL (60g/l), Remazol RedX3BS (20g/l), Water/Oil Repellent (30 g/l). TCAA (2g/l); Wet Pick up (70%), Drying at 100°C/ 3min, Curing at 160°C/ 3min.

3.6. Bifunctional Finishing and Dyeing Properties

W/O repellent and easy-care function as well as the dyeing properties of the simultaneously dyed / finished fabric samples and k/s as well as overall fastness properties are given in Table 6. On the other hand, the impact of the type of reactive dye on the extent of functional finishing and dyeing of fabric samples is also shown in this Table. For a given finishing and dyeing conditions, it is seen that; I) type of reactive dye has practically effect on the extent of inter-crosslinking of polyamide-6/cotton chains, expressed as CRA as well as the W/O repellent rating, II) the enhancement in the CRA, W/O repellent as well as overall fastness properties of the used substrate reflects the positive impact of the decomposition of trichloroacetic acid during curing via acidic, neutral and alkaline medium, III) fastness to washing, rubbing and perspiration of the obtained reactive dyeings are in the range of 4 to 4-5.

4. CONCLUSIONS

Multi-functional finishing/reactive dyeing of polyamide-6 /cotton blends (50/50) in one stage via trichloroacetic acid (TCAA) as a catalyst has been established for imparting a specific functional properties to the above substrate.

Best results of k/s values, crease recovery angle (CRA), handle, water-oil repellency and overall fastness properties were obtained with 2 g/l of trichlroacetic acid.

The extent of bifunctional finishing/dyeing and performance properties of polyamide-6 / cotton were achieved via 60 g/l of knittex® FEL as a finishing agent.

The best multifunctional finishing/dyeing properties i.e., handle and k/s values as well as over-all fastness properties were achieved via W/O repellent agent at the concentration of 30 g/l.

The optimal curing temperature that brings about a significant improvement in the present maltifunctional finishing/dyeing properties was 160°C.

W/O repellent ratings, CRA, K/S and other functional properties attained maximum values at the optimum curing time used i.e., 3 minutes, and prolonging it up to 5 minutes has no effect.

The obtained data revealed that very good results for the over-all fastness properties of the bifunctional/dyeing polyamide-6 /cotton blends with different remazol dyes were determined at the previous optimum conditions, and they are in the range of 4 to 4-5.

REFERENCES

1. D. Hegemann, M. M. Hassain D. Balazs, Nanostructured plasma coatings to obtain multifunctional surfaces, Prog. Org. Coat. 58 (2007) 237-240.

2. L. H. Lin. C. C. Wang. C. W. Chen. K. M. Chen, Water-repellent and antibacterial activities of plasma-treated cleavable silicone surfactants on nylon fabrics surf. Coatings technical 201 (2002) 674-678.

3. S. Lee. J. S. G. Cho, Antimicrobial and blood repellent finished for cotton and nonwoven fabrics based on chitosan and fluoropolymers, Textile Res. J. 69 (1999) 104-112.

4. P. De. M. D. Sankhe S. S. Chawdhari, M. R. Mathur, UV-resist water- repellent breathable fabric as protective textiles, J. Ind. Textile 34 (2005) 209-222.

5. L. Benisek P. C. Craven Machine-washable, water and oil repellent, flame retardant wool, textile. Res. J. 12 (1980) 705-710.

6. M. M. Millard K.S. Lee. AE. Pavlath, Continuous low-temperature discharge treatment of wool yarn: a process for making wool yarn oil repellent and shrink resistant, Textile Res. J. 42 (1972) 307-313.

7. K. D. Veatch and B. M. Gatewood, AATCC Review, 2, 47-51 (2002).

8. N. A. Ibrahim, R. Refaie and A. F. Ahmed, J. Industrial Textiles, 40, 65-83 (2010).

9. P. Bajaj, "Finishing of Textile Materials", J. Appl. Polymer Sci., 83, 631-659 (2002).

10. Holme, Coloration Technologies, 123, 59-73 (2007).

11. Hebeish and N. A. Ibrahim, Colourage Annual, 54, (4), 41-55 (2007).

12. P. Martel, M. Weltrowski, D. Ruffin and M. Morecellet, J. Appl. Polym. Sci., 83, 1449-1456 (2002).

13. P. Horridge and S. Shan, American Dyestuff Reporter, (7), 38 (1994).

14. Intersectional Technical Paper Competition, Textile Chemist and Colorist 26, (11), 29 (1994).

15. Imada, K., Yoshida, M. Text. Chem. Color 22 (11), 18 (1990).

16. Cardamone, J. M., Bao, G Marmer, W. N. Text. Chem. Color 29 (9), 30 (1997).

17. Cardamone, J. M., Marmer, W. N., Lambert, A. H., Bardy. J. B. Text. Chem. Color 28 (11), 19 (1996).

18. Cardamone, J. M., Bao A., Frances, A; Marmer, W. N. Am. Dyest Rept. 86 (1), 41-47 (1997).

19. Park. J., Shore, J. Text. Chem. Color., Amer. Dyest Rep., 32 (1), 46 (2000).

20. Osterloh, Textile Praxis, 26 (1971) 164.

21. Givse and Stapleton, J. S. D. 91 (1975) 223.

22. Feess, Melhiand Textilber, 51 (1970) 197.

23. Kaiser, Chemiefasern/ Textilindustrie, 26, 128 (1976) E63.

24. Kubelka, P. Munk, F. Texhn. Z. Physik 1931, 12, 593.

25. H. Peterson "Chemical After-treatment of Textiles" Edited by H. Mark, H. S. Wooding and S. M. Atlas, John Wiley & Sons, Inc., (1971).

26. Hebeish, E. A. Allam, A. A. Hamsa and M. Kamel., Kolorisztikat Ertesito, 11.12, 308 (1976).

27. W. K. Lee, and K. M. Sin, Textile Asia, 22 (12), 86 (1991).

28. P. Xiaojum, W. Jitao, and S. Jie, J. S. D. C., 109, 159 (1993).

29. N. A. Ibrahim, A. Amr, B. M. Eid and Z. M. El Sayed, "Innovative Multifunctional Treatments of Ligno. Cellulosic Jute Fabric" Carbohydrate Polymers, 82, 1198-1204 (2010).

30. M. H. Abo Shosha. Z. H. El-Hilw, A. A. Aly and A. Amer, "New Textile Water Repellent Based on Reaction of Toluene 2,4 Düsocyanate with Stearyl Alcohol" AATCC Review, 9, 38-42 (2009).

31. Abo. Shosha, M. H., Ibrahim, N.A. "Water soluble polyurethane as finishing additives is easy care finishing of cotton fabrics, Sci. International (Lahore), 1992, 4 (2), 163.

32. Haggag, K, Ebrahim, N. A. J. Text. 1991, 32. 9.

33. Adams, N. E. Detergent action and its relation to wetting and emulsification. J. Soc. Dyers and Color. 1937, 53, 121.

3

CHAPTER

Investigation into the Effect of Extended Laundering on the KES-F Mechanical Properties of an Easy Care Treated Cotton Fabric

Liberato Venant Haule

College of Engineering and Technology, University of Dar es Salaam, Dar es Salaam, Tanza

ABSTRACT

The effect of extended laundering on cotton fabric treated with Dimethylol dihydroxyethyleneurea (DMDHEU) easy care finish was investigated and the fabric characterised by crease recovery performance and the Kawabata Evaluation System for Fabrics (KES-F). The KES-F results indicated that the mechanical handle properties of the DMDHEU treated cotton fabrics were affected by both the levels of application of the DMDHEU easy care finishes and the stress relaxation of the fabrics in aqueous conditions.

Keywords: DMDHEU, Extended Laundering, KES-F, Cotton Fabrics, Easy Care

1. INTRODUCTION

Dimethylol dihydroxyethyleneurea (DMDHEU) is applied to cotton fabrics in order to impart easy care properties such as crease resistance/recovery and dimensional stability. However generally the crease recovery performance is improved at the expense of strength loss and increased shear and bending rigidity [1] - [7] . Some of the detrimental effects of the easy care finish on the mechanical properties of the cotton fabrics can be reduced by incorporating softeners in the easy care formulation [8] [9] . In order to overcome the disadvantages of the DMDHEU easy care finishes, extensive research has focused on the polycarboxylic acids based easy care finishes and improved strength recovery has been reported [10] - [20] . However, DMDHEU still dominates the textile finishing due to its outstanding durability and low cost which attracts more attention.

Previous studies on the wash durability of the DMDHEU treated cotton fabrics as assessed by the crease recovery angle (CRA) performance and surface and bulk analysis indicated that the DMDHEU easy care finish applied to cotton garment is durable almost for the entire life of the garment [21] . An investigation on the wash durability of the DMDHEU on domestic laundering machines and wet cleaning conditions indicated that the easy care finish is less durable on domestic laundering machine than wet cleaning conditions due to the aggressiveness of the former. During the study, the hand (tactile) properties of the fabrics were only evaluated based on the coefficient of friction and geometrical roughness using the KES-F system. The surface friction of the investigated fabrics was found to be higher for the domestic laundering than wet cleaning conditions due to fibrillation tendency of the substrates in former. The effects of DMDHEU concentrations on the hand properties were not considered [22] .

In current study, the effects of extended laundering of DMDHEU treated cotton fabric on the fabric's mechanical properties as determined by the Kawabata Evaluation System for Fabrics (KES-F) were investigated.

The KES-F system was developed in Japan by a group led by Professor S. Kawabata in 1968 and came into operation 10 years later [23] - [26] . Originally the system was developed for objective evaluation of fabric "handle", and was also commonly used for product/process development, process control and optimisation. The KES-F technique measures the fabric mechanical and surface properties, Table 1, at load levels typical of normal human handling and end-user applications and the objective measurements are correlated to subjective handle properties of the material. The acquired handle properties such as fullness, smoothness and stiffness can objectively be compared and adjusted for quality and performance control purposes. Fabric making up performance can be controlled by predicting problems such as seam pucker from the KES-F fabric "fingerprint" and fabric drape and other end-use characteristics can also be predicted from the objective measurements.

Previous work suggested that the KES-F tests may be used for investigating the mechanical changes of fabrics during wet processing and it was reported that fabric laundering reduces stiffness and increase smoothness, softness and fullness of processed cotton fabrics [26] . The KES-F test was used to assess the effects of fabric raisings on comfort, hand and mechanical properties of artificial suede. The increased number of raisings caused the suede to have a soft and smooth surface and improved perception of luxuriousness [27] . The hand properties of woven and knitted fabrics were investigated using the KES-F and frictional analyser tests and the KES-F results indicate that there is a good correlation between friction and tactile properties for knitted fabrics but not for woven fabrics [28] . Furthermore the Kawabata system was applied on investigation on the effects of fabric movements in front-loading washers on hand properties. The study established a relationship between fabric movements and hand properties and the relationship could be used by the washing machine manufacturers on design of washing programmes of various types of garments [29] . The Kawabata system has also been tested and recommended for the paper industry for the determination of objective mechanical properties which can be used to translate the subjective handle properties such as softness/smoothness [30] . KES-F was used for investigating the performance of the TiO_2 as an easy care finishes for cotton material [31] .

Table 1. Mechanical parameters measured by the KES-F system.

Parameter	Symbol	Units
Tensile		
Tensile linearity	LT	-
Tensile energy per unit area	WT	gf.cm/cm^2
Resilience	RT	%
Extension at specific load	EMT	%
Shear		
Shear stiffness	G	gf/cm.deg
Hysteresis at 0.5°	2HG	gf/cm
Hysteresis at 5°	2HG5	gf/cm
Bending		
Bending stiffness	B	gf.cm^2/cm
Hysteresis width	2HB	gf.cm/cm
Compressional		
Compressional linearity	LC	-
Energy of compression	WC	gf.cm/cm^2
Resilience	RC	%
Compressibility	C	%
Thickness and weight		
Thickness under a force of 0.5 gf/cm^2	To	mm
Weight	W	mg/cm^2
Thickness under a force of 50 gf/cm^2	Tm	mm
Surface		
Coefficient of friction	MIU	-
Mean deviation of MIU	MMD	-
Geometrical roughness	SMD	μm

In this paper the effects of extended laundering, with perborate and a non-perborate detergent, on the KES-F handle properties of the DMDHEU treated cotton fabrics was investigated.

2. METHODOLOGY

2.1. Materials

Scoured, bleached, 100% plain woven cotton fabric, 152 g/m^2 was supplied by Phoenix Calico, UK. Fixapret CP New (dimethylol dihydroxyethyleneurea, DMDHEU) was supplied by Dystar, UK. Magnesium chloride was supplied by Fisher Scientific, UK. The ECE non-phosphate detergent, ECE phosphate detergent and Tetra Acetyl Ethylene Diamine (TAED, 92% active) were purchased from the Society of Dyers and Colourists (SDC), Bradford , UK . Sodium Perborate Tetrahydrate (97%) active was purchased from Aldrich Chemicals Ltd., UK .

2.2. Treatment Conditions

Aqueous solutions of DMDHEU, 60, 100 and 140 g/L, and magnesium chloride , 12 g/L, 20 g/L and 28 g/L, respectively, were padded onto the cotton fabric at 80% wet pick up. The magnesium chloride was included in the padding solution as a catalyst during curing. The padded fabrics were oven dried for 2.5 minutes at 100°C with subsequent curing for 4 minutes at 150°C. The cured fabrics were rinsed in warm water to remove any finish residue and then air dried for further tests.

Untreated and DMDHEU treated cotton fabrics were repeatedly laundered, up to 10 cycles, in a Wascator FOM 71 MP washing machine using the 5A wash cycle programme. The machine complies with British Standards for domestic washing and drying testing procedures, BS EN ISO 6330:2001 + A1:2009. The washing was performed with 12.5 g of ECE non phosphate detergent in the presence and absence of 30 g sodium perborate and 4 g of TAED per wash.

2.3. Crease Recovery Angle (CRA) Analysis

The crease recovery of the fabrics was determined using the British Standard test method, BS EN 22313:1992 [32] , with at least ten replicates performed in both and the mean for both warp and weft (W+F) reported.

2.4. Kawabata System Evaluation for Fabrics (KES-F) Analysis

The samples, 20 cm × 20 cm, were conditioned at 65% R.H. and 20°C for 24 hrs prior to testing. In this work only the tensile, shear, bending and compressional properties were analysed. The results presented were the average of three measurements taken along the warp and weft directions of the tested fabrics.

3. RESULTS AND DISCUSSION

Examination of the CRA performance of the cotton fabrics treated with DMDHEU easy care finish indicates the CRA performance increased with an increase in application of DMDHEU, Figure 1. When the DMDHEU treated cotton fabric was laundered both with/without perborate there were marginal decreases in the CRA

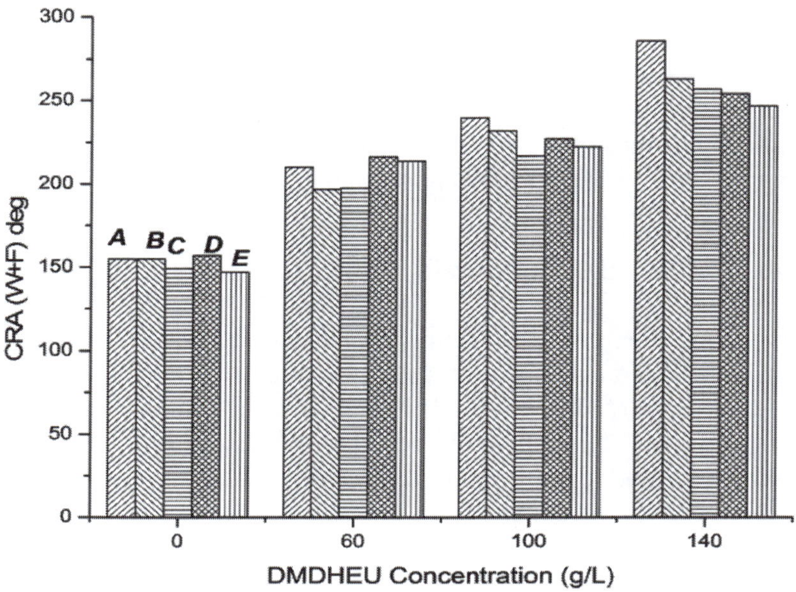

Figure 1. Effect of repeated washes, A—unwashed, B—one wash with perborate, C—ten washes with peborate, D—one wash without perborate, E—ten washes without perborate on the CRA performances. (W + F) = warp + weft.

performance in the concentration range of 60 - 100 g/L DMDHEU, but the decrease did not reach the level of the control cotton fabric (0 g/L DMDHEU). This observation agrees with the previously reported information that the DMDHEU is durable even after extended laundering [21] . While the decrease in CRA performance with the number of washes was relatively small in the 0 - 100 g/L range, the decrease was more obvious at the 140 g/L DMDHEU application level, Figure 1. The nature of this phenomenon was discussed in detail in a previous study where the type of cellulosic crosslink introduced at varying levels of DMDHEU treatment was investigated.

Figure 2 presents the effect of extended washing on the extensibility (EMT) of the untreated and DMDHEU treated cotton fabric, as measured by the KES-F system. Treatment of the cotton fabric with increasing concentration (60 - 100 g/L) DMDHEU produced an increase in the fabric extensibility. In contrast application of DMDHEU at 140 g/L decreased the fabric extensibility below that of the untreated cotton. Fabric extensibility depends on the flexibility of the cellulose chain molecules to move. Application of DMDHEU crosslinks the cellulose chain molecules and thus restricts the mobility of the chains, while the aqueous padding of the fabric results in fabric relaxation and increased the fabric extensibility at 60 - 100 g/L levels. At 140 g/L DMDHEU the fabric relaxation effect cannot overcome the crosslinking effect, thus a measurable decrease in extensibility was observed. Subsequent Wascator washing relaxes the fabric, breaks bonds and increased EMT. Perborate washing caused greater oxidative damage and hence reduced EMT marginally less relative to the non-perborate washing.

The effect of repeated laundering on untreated fabric was to increase the fabric thickness and compressibility, Figure 3 and Figure 4. This increase was probably due to the fibrillation of the fibres and the formation of hairier and bulkier fabrics and stress relaxation of the fabric during washing. The effect of initial washing with perborate on the thickness of the fabrics treated with DMDHEU concentration was to reduce the fabric thickness T_m and T_o, Figure 3 and Figure 4. This reduction in fabric thickness with the perborate wash treatment was probably due to its oxidative degradative effect preferentially removing embrittled surface fibres. Subsequently further fibres are able to become disengaged with the yarn structure, create a surface layer of protruding fibres and fabric thickness increased.

The effect of increasing DMDHEU application level was overall to increase the fabric rigidity above that of the untreated fabric, Figure 5. However with the 60 - 100 g/L applications the bending rigidity actually decreased due to the fabric relaxation effect outweighing the crosslinking effect of the formaldehyde-based crosslinker. Application of DMDHEU at the 140 g/L level increased the bending rigidity above the untreated cotton value due to the crosslinking. In general subsequent laundering relaxed the fabric, broke internal bonds and decreased the bending rigidity. Little difference between the perborate wash and non-perborate washings was observed.

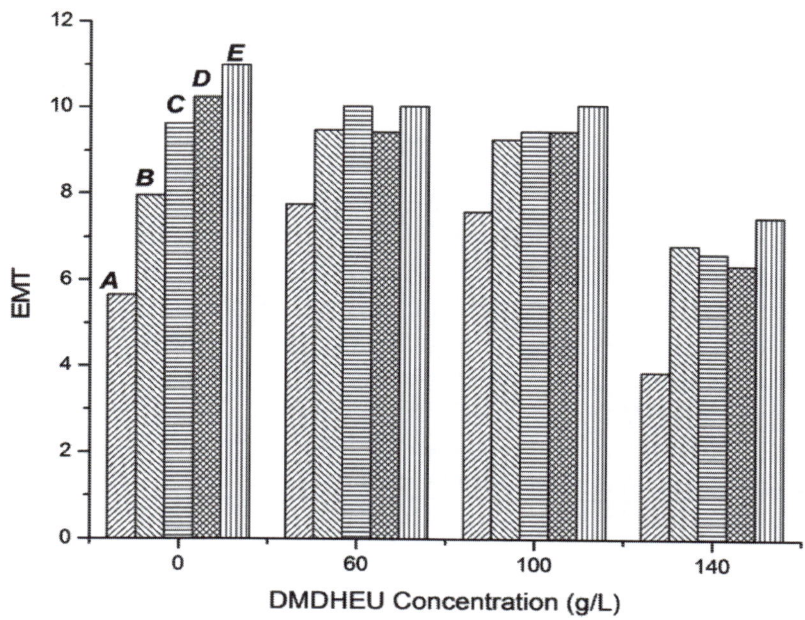

Figure 2. Effect of repeated washes, A—unwashed, B—one wash with perborate, C—ten washes with peborate, D—one wash without perborate, E—ten washes without perborate on the extension.

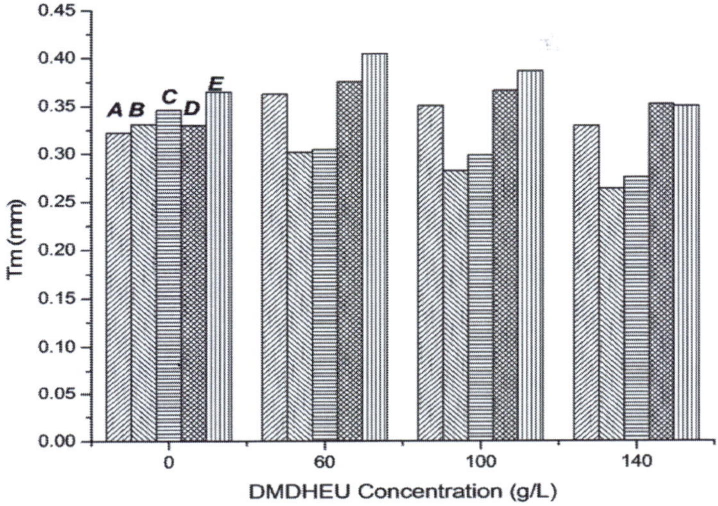

Figure 3. Effect of repeated washes, A—unwashed, B—one wash with perborate, C—ten washes with peborate, D—one wash without perborate, E—ten washes without perborate on the thickness (T_m) for cotton fabric.

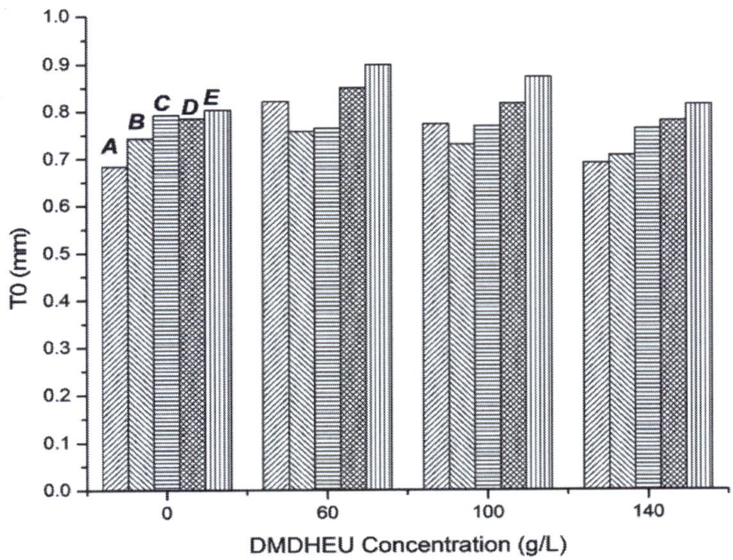

Figure 4. Effect of repeated washes, A—unwashed, B—one wash with perborate, C—ten washes with peborate, D—one wash without perborate, E—ten washes without perborate, on the thickness (T_0).

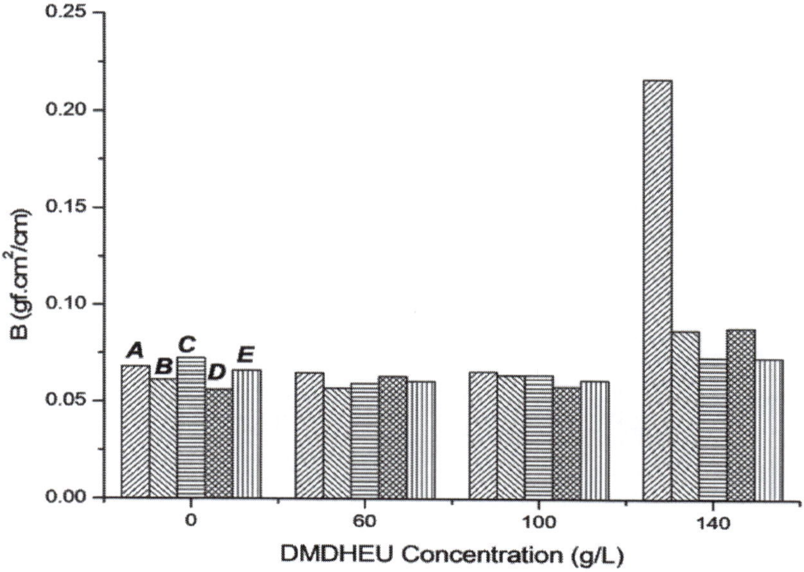

Figure 5. Effect of repeated washes, A—unwashed, B—one wash with perborate,C—ten washes with peborate, D—one wash without perborate, E—ten washes without perborate on the bending rigidity.

Figure 6 illustrates the effect of increasing DMDHEU treatment levels of cotton fabric and extended washing on the shear stiffness (G) of the cotton fabric as measured by the KES-F system. Application of DMDHEU at 140 g/L increased G due to crosslinking of the cellulose chain molecules within the fibre/yarn. Below this level the effect of aqueous padding relaxing the fabric predominated and decreased G. At 60 - 100 g/L the crosslinking was counteracted by this relaxation. Subsequent Wascator washing relaxed the fabric, broke bonds and decreased G for the 140 g/L application and untreated cotton fabrics. For other application levels washing increased G and again little difference between perborate wash and non-perborate washing was observed.

Figure 7 presents the effect of DMDHEU concentration and extended Wascator washing on the shear hysteresis (2HG5) of the cotton fabric as measured by the KES-F system. Application of DMDHEU at 140 g/L increased 2HG5 due to crosslinking of cellulose chains in the fibre. Below this level aqueous padding relaxed the fabric and decreased 2HG5. At 60 - 100 g/l the crosslinking/embrittling effect was counteracted by this relaxation. Subsequent Wascator washing relaxed the fabric, breaks bonds and decreased 2HG5 for the 140 g/L application and untreated fabric. For the other fabric treatments washing increased 2HG5 probably due to fibre surface modification increasing inter-fibre frictional interactions. Little difference between perborate wash and non- perborate washing was observed.

Re-examination of the effect of repeated Wascator washing on the CRA performance, Figure 1, bending properties, Figure 5 and Figure 6 and shear properties Figure 7 indicated that the 140 g/L DMDHEU treated cotton has the relatively largest decrease in CRA performance, bending and shear rigidity properties when compared to 60 - 100 g/L DMDHEU after repeated washing. Previous research has demonstrated that crosslinking of cotton fabrics with DMDHEU

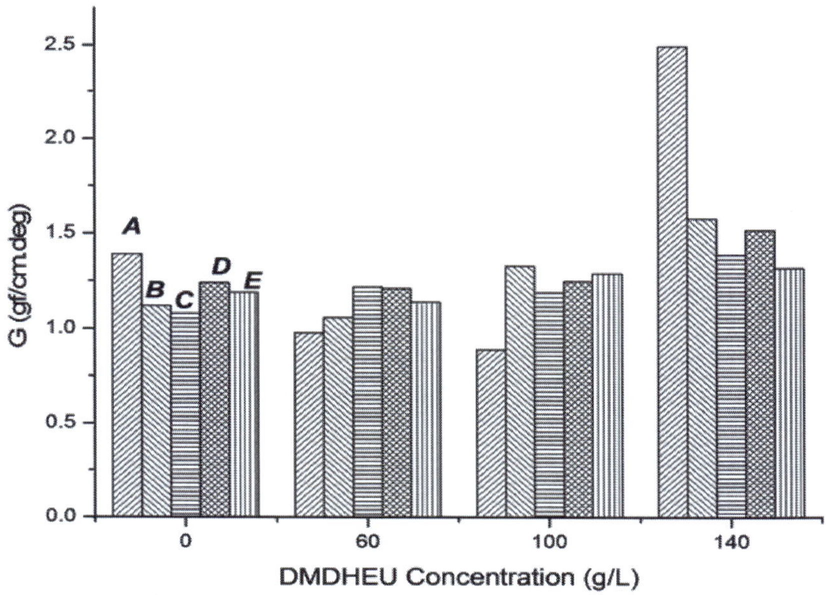

Figure 6. Effect of repeated washes, A—unwashed, B—one wash with perborate, C—ten washes with peborate, D—one wash without perborate, E—ten washes without perborate on the shear rigidity.

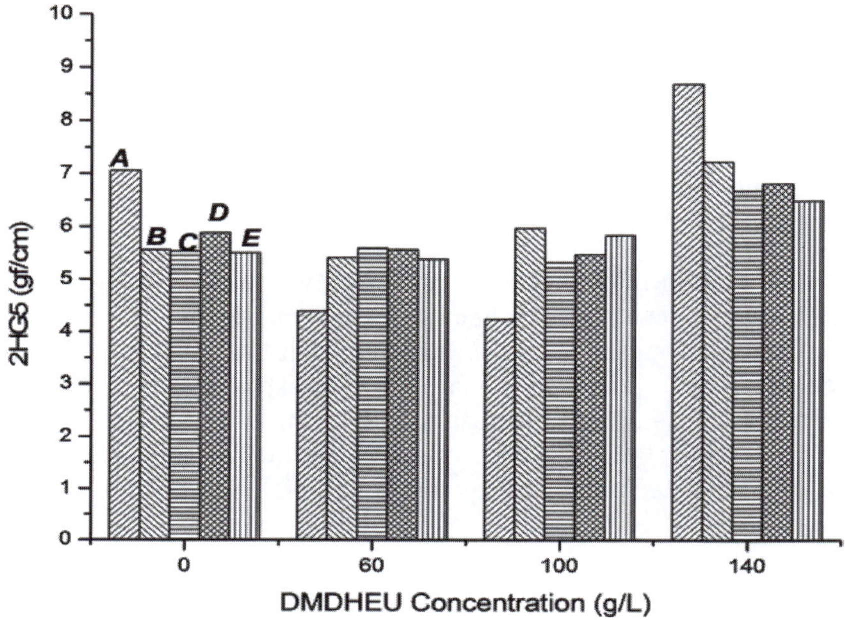

Figure 7. Effect of repeated washes, A—unwashed, B—one wash with perborate, C—ten washes with peborate, D—one wash without perborate, E—ten washes without perborate on the shear hysteresis at 5° shear angle.

concentrations at lower than 140 g/L favours tri- and tetra-functional crosslinks and the crosslinks are stable to extended laundering. However DMDHEU application at 140 g/L is dominated by unstable bi-functional crosslinks [21] . Hence the measurable decrease in CRA performance, bending and shear properties of the cotton 140 g/L DMDHEU treated, washed cotton fabric may be explained by the bi-functional crosslinks which are unstable to repeated Wascator washing and are prevalent at this higher application levels. Therefore the optimal application levels of DMDHEU easy care finish in cotton fabrics is anticipated to be in 60 - 100 g/L.

4. CONCLUSION

Cotton fabrics were treated with DMDHEU at increasing concentration and the effect of extended laundering on the cotton fabric treated with DMDHEU easy care finish was investigated. The presence and durability of the easy care finish on the cotton fabric was monitored using the CRA performance. The CRA performance test results indicated that DMDHEU treatment at (0 - 100 g/L) was stable to extended laundering conditions, however the 140 g/L treatment was significantly affected by the extended laundering conditions and can be explained by the nature of crosslinking between cellulose polymer and DMDHEU at varying concentrations. The KES-F results indicated that the handle properties of the DMDHEU treated cotton fabrics were affected by both the level of application of the DMDHEU easy care finish and the stress relaxation of the fabrics in aqueous conditions. It was evident that the presence of perborate in the wash formulation had relatively little effect on the fabric handle properties above that observed with laundering with the non-perborate formulation.

REFERENCES

1. Grant, J.N., Orr, R.S., Tsoi, R.H. and Weiss, L.C. (1966) Strain-Recovery Properties of Wash-Wear-Treated Cotton Fabrics. Textile Research Journal, 36, 642-650.

2. Gilbert, S.M. and Smith, B.F. (1970) Degradation Study of Some Formaldehyde-Modified, Celluloses. Textile Research Journal, 40, 720-727

3. Hassenboflfr, C.B.J.R., Harper, R.J. and Grant, J.N. (1971) Tensile, Shear, and Recovery Properties of Cotton Fibers and Fabrics Treated to Impart Differences in Wet and Dry Wrinkle Recoveries. Textile Research Journal, 41, 219-225

4. Murphy, A.L., Margavio, M.F. and Welch, C.M. (1971) All-Cotton Durable-Press Fabrics of High Strength from Slack-Mercerized, Partially Restretched Yarn: Evidence for the Separate Origins of Strength and Wrinkle-Resistance Effects

in Conventional Cross-Linking Treatments for Cotton. Textile Research Journal, 41, 22-3

5. Grant, J.N., Andrews, F.R. and Weiss, L.C. (1971) Physical Properties of Cotton Treated by Pad-Dry-Cure, Mild-Cure, Poly-Set, and Wet-Fix Processes. Textile Research Journal, 41, 44-51.

6. Raheel, M. and Lien, M.D. (1982) Modifying Wear Life of All-Cotton Fabrics: Part I: Liquid Ammonia Treatment and Durable Press Finish. Textile Research Journal, 52, 493-503.

7. Jones, B.W., Turner, J.D. and Luparello, D.O. (1980) Preparing and Finishing Durable-Press Cotton Fabrics1. Textile Research Journal, 50, 165-177.

8. Hashem, M., Ibrahim, N.A., El-Shafei, A., Refaie, R. and Hauser, P. (2009) An Eco-Friendly-Novel Approach for Attaining Wrinkle-Free/Soft-Hand Cotton Fabric. Carbohydrate Polymers, 78, 690-703.

9. Ibrahim, N.A., Abo-Shosha, M.H., Fahmy, H.M., El-Sayed, Z.M. and Hebeish, A.A. (2008) Hybrids for Finishing Cotton Fabric with Durable Handle Performance. Journal of Materials Processing Technology, 200, 385-389.

10. Yang, C.Q. (2001) FTIR Spectroscopy Study of Ester Crosslinking of Cotton Cellulose Catalyzed by Sodium Hypophosphite. Textile Research Journal, 71, 201-206.

11. Yang, C.Q. (1991) FT-IR Spectroscopy Study of the Ester Crosslinking Mechanism of Cotton Cellulose. Textile Research Journal, 61, 433-440

12. Yang, C.Q. (1991) Characterizing Ester Crosslinkages in Cotton Cellulose with FT-IR Photoacoustic Spectroscopy 1. Textile Research Journal, 61, 298-305

13. Yang, C.Q. and Andrews, B.A.K. (1991) Infrared Spectroscopic Studies of the Nonformaldehyde Durable Press Finishing of Cotton Fabrics by Use of Polycarboxylic Acids. Journal of Applied Polymer Science, 43, 1609-1616.

14. Yang, C.Q., Chen, D., Guan, J. and He, Q. (2010) Cross-Linking Cotton Cellulose by the Combination of Maleic Acid and Sodium Hypophosphite. 1. Fabric Wrinkle Resistance. Industrial & Engineering Chemistry Research, 49, 8325-8332.

15. Yang, C.Q. and Dengjin, W. (2000) Evaluating Ester Crosslinking of Cotton Fabric by a Polycarboxylic Acid Using Acid-Base Titration. Textile Research Journal, 70, 615-620.

16. Yang, C.Q., Hu, C. and Lickfield, G.C. (2003) Crosslinking Cotton with Poly (Itaconic Acid) and in Situ Polymerization of Itaconic Acid: Fabric Mechanical Strength Retention. Journal of Applied Polymer Science, 87, 2023-2030.

17. Yang, C.Q., Lan, X., Shiqi, L. and Yanqiu, J. (1998) Nonformaldehyde Durable Press Finishing of Cotton Fabrics by Combining Citric Acid with Polymers of Maleic Acid. Textile Research Journal, 68, 457-464.

18. Yun, L. and Yang, C.Q. (1999) Fabric Yellowing Caused by Citric Acid as a Crosslinking Agent for Cotton. Textile Research Journal, 69, 685-690.

19. Weilin, X. and Shyr, T.-W. (2000) Durable Press Finishing of Cotton Fabrics by Polycarboxylic Acids After Graft Copolymerization with Hydroxyethyl Methacrylate. Textile Research Journal, 70, 8-10.

20. Xu, W. and Shyr, T. (2001) Applying a Nonformaldehyde Crosslinking Agent to Improve the Washing Durability of Fabric Water Repellency. Textile Research Journal, 71, 751-754.

21. Haule, L.V., Rigout, M., Carr, C.M. and Jones, C.C. (2012) Surface and Bulk Chemical Analysis of the Durability of an Easy Care Finish on Cotton. Cellulose, 19, 1023-1030

22. Rijavec, T., Jevsnik, S., Soljacic, I., Tomljenovic, A., Visic, K. and Pusic, T. (2015) Physical and Chemical Effects of Washing and Wet Cleaning on Durable Press Finished Cotton Fabrics. Textile Research Journal, 85, 632-645

23. Shenai, V.A. (1989) Fundamental Concepts of Fabric Hand. Part 1. Textile Dyer and Printer, 22, 19-26.

24. Shenai, V.A. (1989) Fundamental Concepts of Fabric Hand. Part 2. Textile Dyer and Printer, 22, 15-22.

25. Fritz, A. (1972) New Way to Measure Fabric Hand. Textile Asia, 23, 69-72.

26. Carr, C.M. (1988) Investigation of the Effect of Water Repellent Finishes on the Mechanical Properties of Textiles Using the Kawabata Evaluation System for Fabrics. Journal of Industrial Textiles, 18, 106-113.

27. Roh, E.K., Oh, K.W. and Kim, S.H. (2014) Effect of Raising Cycles on Mechanical, Comfort, and Hand Properties of Artificial Suede. Textile Research Journal, 84, 1995-2005.

28. Bertaux, E., Lewandowski, M. and Derler, S. (2007) Relationship between Friction and Tactile Properties for Woven and Knitted Fabrics. Textile Research Journal, 77, 387-396.

29. Park, S., Yun, C., Kim, J. and Park, C.H. (2013) The Effects of the Fabric Properties on Fabric Movement and the Prediction of the Fabric Movements in a Front-Loading Washer. Textile Research Journal, 83, 1201-1212.

30. Parfitt, M., Vickerman, J.C., Mitchell, R., Carr, C.M., Ince, N. and Knight, P. (2003) Surface Analysis of Softened Paper by Time-of-Flight Secondary Ion Mass Spectrometry (ToF-SIMS) and the Kawabata Evaluation System. Journal of Materials Science, 38, 2171-2177.

31. Lam, Y.L., Kan, C.W., Yuen, C.W. and Au, C.H. (2011) Low Stress Mechanical Properties of Plasma-treated Cotton Fabric Subjected to Titanium Dioxide Coating. Textile Research Journal, 81, 1008-1013

32. British Standards Online (1992) Textile Fabrics. Determination of the Recovery from Creasing of a Horizontally Folded Specimen by Measuring the Angle of Recovery: BS EN 22313:1992, ISO 2313:1972.

4

CHAPTER

Effect of Different Dyeing Parameters on Color Strength & Fastness Properties of Cotton-Elastane (CE) and Lyocell-Elastane (LE) Knit Fabric

Md. Abdus Shahid[1], Md. Ismail Hossain[2], Delwar Hossain[1], Ayub Ali[3]

[1]*Dhaka University of Engineering & Technology (DUET), Department of Textile Engineering, Gazipur, Bangladesh*

[2]*University of Malaya (UM), Department of Mechanical Engineering, Kuala Lumpur, Malaysia*

[3]*BGMEA University of Fashion and Technology (BUFT), Department of Apparel Manufacture & Technology, Dhaka, Bangladesh*

ABSTRACT

Color fastness and color strength are very crucial facts in case of colored textile material which are intimately related to various dyeing parameters viz. concentration of dye, salt, alkali and dyeing temperature. In the present work, the effect of concentration of reactive dye (1%, 3%, 5%, 7% & 9%), salt (20g/l, 30g/l & 40g/l) and alkali (5g/l, 8g/l & 10g/l), and the effect of dyeing temperature (40°C, 60°C & 80°C) on color fastness and color strength properties of cotton-elastane (CE) and lyocell-elastane (LE) fabrics were studied remaining the constant of dyeing duration time at 45 minutes. The results were compared for variation in fastness values and color strength under the same conditions as being employed in textile industry. Some of the factors behave linearly and some are not. Result reflects that the color fastness values to dry and wet rubbing are more at low dye concentration, low salt and alkali concentration and at low temperatures for both cotton-elastane and lyocell-elastane fabrics which reaches up to 4-5 to 5. But in case of washing fastness property, both fabrics reflect a fixed value i.e.4-5 irrespective of dye concentration, salt and alkali concentration, dyeing time and temperature. On the other hand, the color strength results evinced that temperature has more influence on LE than CE for acquiring higher color strength. LE fabric exhibited better dye ability with higher color uptake compared to CE fabric. In addition the concentrations of salt and alkali have almost linear proportional effect on both CE and LE within a range.

Keywords: Cotton-Elastane, Lyocell-Elastene, Color Fastness, Color Strength, Reactive Dye, Salt, Alkali

1. INTRODUCTION

The generations of regenerated cellulose fibers, such as viscose, modal and lyocell fibers are among the most important fibers from the point of textile and environmental aspects due to the natural structures and properties [1]. Lyocell is the generic name for a regenerated cellulosic fiber obtained by spinning of dissolved wood pulp in an organic solvent. These regenerated cellulosic fibers are produced by a more environmentally friendly procedure where cellulose is dissolved in the organic solvent N-methyl morpholine-N-oxide (NMMO). Regeneratedcellulosic yarns with various chemical reagents had been the subject of research for several years for NMMO or lyocell fibers, produced by the amine oxide process [2, 3] are solvent spun. More than 99% of the solvent is recycled within the process, making the fiber production ecofriendly. Very small quantities of NMMO emitted via waste water are readily degraded in the biological waste water treatment plants. NMMO with its favorable toxicological properties has been found to be less toxic than ethanol. The solvent spinning technique so adopted is an environmentally responsible process [4]. This manufactured fiber shows round cross-sections and a smooth surface. Their high degree of crystallinity, molecular orientation and higher molecular weight causes special properties such as high wet fiber strength and comparatively low elongation, leading to the remarkable dimensional stability of fabrics produced from them. Lyocell fiber's degree of crystallinity is 16 % higher than that of modal fibers and much higher (43%) compared to viscose fibers. The fibrils in lyocell fibers are thinner than in modaland viscose fibers and molecular orientation is nearly the same (orientation function of lyocell fibers is 3% higher) as in modal fibers but much higher compared to viscose standard fibers (18%). Lenz and Schurz [5] determined the size of crystalline regions of regenerated cellulose fibers: length 12 - 14 nm, width 8 - 10 nm, and thickness 3 - 4 nm. The broad plane of the platelet-shaped crystallites is situated parallel to the 101 - lattice plane. The crystallites form strands with a length of 150 nm - 550 nm. They are partly bundled up in clusters with diameters of 30 - 60 nm, partly separated by less dense spacing regions. Since lyocell is cellulose based fiber and therefore can be dyed with any class of dyestuff suitable for other cellulosic fiber. On the other hand cotton fibers are widely applied in textile industry due to its excellent properties of hygroscopicity, air permeability, biodegradability, no static electricity, etc. The dyeing of these fibers are generally done with reactive dyes due to its brilliancy, variety of hue, high wet fastness, convenient usage and high applicability. These reactive dyes contain a reactive group, either a halo-heterocyclic or an activated double bond, that, when applied to a fiber in an alkaline dye bath, forms a chemical bond with hydroxyl group on the cellulosic fiber [6]. In recent years, reactive dyes maintain the largest annual consumption in the world among the dyes used for which establishes its important status in the dye manufacture industry. But some problems, such as low dye utilization, large amount of electrolyte used and high volume of wastewater discharged, always exist in the application of reactive dyes. The dyeing of one kilogram of cotton with reactive dyes demands from 70 to 150 litre water, 0.6-0.8 Kg NaC1 and from 30 to 60 g dyestuffs [7]. Spandex/elastane is an elastomeric which is always blended with other natural and synthetic fibers such as cotton, wool, silk, and linen. Different types of spandex material exist on the markets with different counts, this impose engineering the use of every type at different spun materials and fabric construction. The percentage of spandex used in the knitting industry depends on the fabric stretch required and fabric properties [8]. The reactive dyes have higher natural affinity for LE fabric than for CE fabric because of its structure as a result it shows higher K/S value in comparison with CE fabric. Hence, reactive dyes with good migration properties will be best suited for LE fabric providing level dyeing effect. Thus, the electrolyte recommendations can also be significantly reduced, as compared to the dyeing to the same depth of shade on CE fabric. Consequently, there is reduction in chemical cost, handling and total dissolved solids. Lower electrolyte also facilitates quicker removal of unfixed hydrolysed dye from the fiber in washing stage after dyeing, thus reducing the process cycle time and hence the cost [9]. Traditionally disperse dye was used to dye spandex/ superfine polyester fabric and some auxiliaries were also approached for dyeing of elastic fiber with disperse dye [10, 11]. Later on reactive dyes are in fact seen as the major dye class used on LE fabric and CE fabric. However, it is revealed that cotton with less elastane percentage shows better fastness property in case of reactive dyeing [12]. There are several studies done on the optimization of process parameter for obtaining improved dyeing behavior. The process parameters have a great influence on the dyeing performance [13, 14]. Some pretreatment process and treatments give better result in colorfastness property of polyester-spandex fabric [15]. In case of dyeing the CE fabric, ways of getting better color fastness property has been explained by using auxiliaries (HYS-422 auxiliaries) [16]. However, the effect of variation of different conditions (temperature and dye concentration) and auxiliaries (salt and alkali) have not been found during researching by the authors. Furthermore another most eco-friendly fiber LE fabric dyed with reactive dye and the influence the process parameters along with their comparison with CE in respect to color fastness and color strength requires clarification. The aim of this study was to investigate and compare the effect of temperature, amount of alkali, amount of salt and dye concentration ondyability characteristics especially on color fastness (rubbing and wash) and color strength of CE and LEknit fabric.

2. MATERIALS AND METHODS

2.1. Procedure of CE and LE Knit Fabric Manufacturing

In this investigation, single jersey cotton-elastane (CE) and lyocell-elastane (LE) knit fabrics were manufactured by single jersey knitting machine having 32 inches diameter, 24 gauges (needles/inch) and 104 yarn feeders (Brand: Fukuhara, Model: MXC-S3.2RE, Origin: Japan). CE knit fabric was made from 95% cotton & 5% elastane yarn and LE grey fabric was made from 95% lyocell & 5% elastene yarn. The yarn count was 32^s/1for cotton, 18 tex for lyocell, 30 denier for lycra. 100% cotton yarn was sourced from Square Spinning mills Ltd. Bangladesh, lyocellyarn (Brand: Tencell) from Austria, and elastence yarn (Brand: Texlon) from Korea.

2.2. Pre-treatment & Dyeing Procedure

Both types of grey fabrics were scoured and bleached using single bath. Recipe for CE fabric was NaOH @1.2 g/l, H_20_2@2.5 g/l at 98°C for 60 minutes during scouring & bleaching. On the other hand, recipe for LE fabric was Na_2CO_3@4 g/l, H_20_2@2.5 g/l at 80°C for 60 minutes. Single bath scouring and bleaching process was done in a laboratory dyeing machine (Brand: Starlet, Model: DL-6000, Origin: Korea) maintaining the material & liquor ratio of 1:6. Average grams per square meter (GSM) of scoured & bleached fabric were 185. All pre-treated CE and LE knit fabric samples were dyed using exhaust dyeing methods with reactive dye (Remazol Blue RR) in a laboratory dyeing machine (Brand: Starlet, Model: DL-6000+, Origin: Korea) according to Table 1 at material and liquor ratio of 1:8. Forty five experiments were carried out altering the dyeing parameters viz. concentration of dye, salt & alkali and temperature remaining the constant of dyeing duration time at 45 minutes that is showed in Table 1. Table 1 shows that five different level of dye concentration namely 1%, 3%, 5%, 7% and 9% with different salt & alkali concentration and dyeing temperature were used for preparing the samples during this study. After dyeing all samples were cold rinsed and then hot washed by washing machine (Brand: LG, Model: WD-1203 CDP, Origin: Korea), at 90°C for 10 minutes. Next, the samples were dried by tumble dryer machine (Brand: SDL, Origin: UK).and conditioned for 2 hours at (65±2) % RH (relative humidity) and (20±2) °C temperature.

Table 1. Experimental design considering the dyeing parameters

Exp. no.	DC %	Dye temp (°C)	Time (min)	SC (g/l)	AC (g/l)
1	1	40	45	20	5
2	1	60	45	20	5
3	1	80	45	20	5
4	3	40	45	20	5
5	3	60	45	20	5
6	3	80	45	20	5
7	5	40	45	20	5
8	5	60	45	20	5
9	5	80	45	20	5
10	7	40	45	20	5
11	7	60	45	20	5
12	7	80	45	20	5
13	9	40	45	20	5
14	9	60	45	20	5
15	9	80	45	20	5
16	1	40	45	30	8
17	1	60	45	30	8
18	1	80	45	30	8
19	3	40	45	30	8
20	3	60	45	30	8
21	3	80	45	30	8
22	5	40	45	30	8
23	5	60	45	30	8
24	5	80	45	30	8
25	7	40	45	30	8
26	7	60	45	30	8
27	7	80	45	30	8
28	9	40	45	30	8
29	9	60	45	30	8
30	9	80	45	30	8
31	1	40	45	40	10
32	1	60	45	40	10
33	1	80	45	40	10
34	3	40	45	40	10
35	3	60	45	40	10
36	3	80	45	40	10
37	5	40	45	40	10
38	5	60	45	40	10
39	5	80	45	40	10
40	7	40	45	40	10
41	7	60	45	40	10
42	7	80	45	40	10
43	9	40	45	40	10
44	9	60	45	40	10
45	9	80	45	40	10

*SC=Salt concentration, AC=Alkali concentration, DC%=Dye concentration %

2.3. Measurement of Color Strength and Color Fastness Properties

Reflectance values of all dyed samples were measured using spectrophotometer (Brand: Datacolor Spectrum, Origin: USA, Model: 650™), in a visible region with wavelength of 650nm. The color strength (K/S) was calculated using the Kubelka-Munk equation:

$$\frac{K}{S} = \frac{(1-R)^2}{2R}$$

Color fastness properties of all dyed samples were determined using the crock meter (Brand: SDL, Origin: UK) and multi-fiber. The color fastness values were assessed using grey scale both for color fastness to rubbing and color fastness to washing. Rubbing fastness properties were examined by applying BS EN ISO-105-X12:2002 method and color fastness to washing were examined applying BS EN ISO 105 CO6:2010 method.

3. RESULT AND DISCUSSION

3.1. Effect of Temperature on Color Strength and Fastness

3.1.1. Effect of Temperature on Color Strength

K/S value in Figure 1(a), Figure 2(a) and Figure 3(a) reflects that dyeing temperature has a great impact on dye uptake for both fabrics. A dramatic change occurs in fiber polymer system when temperature is applied as a medium of energy. This scenario is best viewed from the plotted curves. As the temperature increases, the molecular structure becomes open which facilitate the dye uptake and hence the higher K/S value is obtained. An interesting relation was obtained when the temperature was increased at different level of dye, salt and alkali concentration. In case of cotton-elastane, the results show that K/S value of dyed fabric increased gradually with the increase of temperature from 40°C to 60°C. In fig 3, at 60 °C for cotton-elastane shows optimum dye uptake but the K/S value in the plotted figure shows that at 80°C the dye uptake results getting down which may be due to diffusion of dye from the core of the fiber. At 1%, 3% and 5% the results shows near parallel effect. On 7% dye it shows slightly higher differences but at 9% dye concentration with 60°C and 80°C shows less difference in results whereas at 40°C it shows declined value. Again, for LE fabric samples, the plotted curves show different action. At 1%, 5% and 7% of dye concentration of 40°C and 80°C shows small value difference. But high level of dye uptake was shown at 60°C for LE. In addition, at 9% dye concentration with 80°C the LE fabrics show a highest peak but decreasing rate of dye uptake at 40°C.The reason for this is obvious and indicates of the effects of different structure for cotton and lyocell. Due to higher degree of crystalinity in the polymer structure of CE to LE causes less dye uptake of CE fabrics.

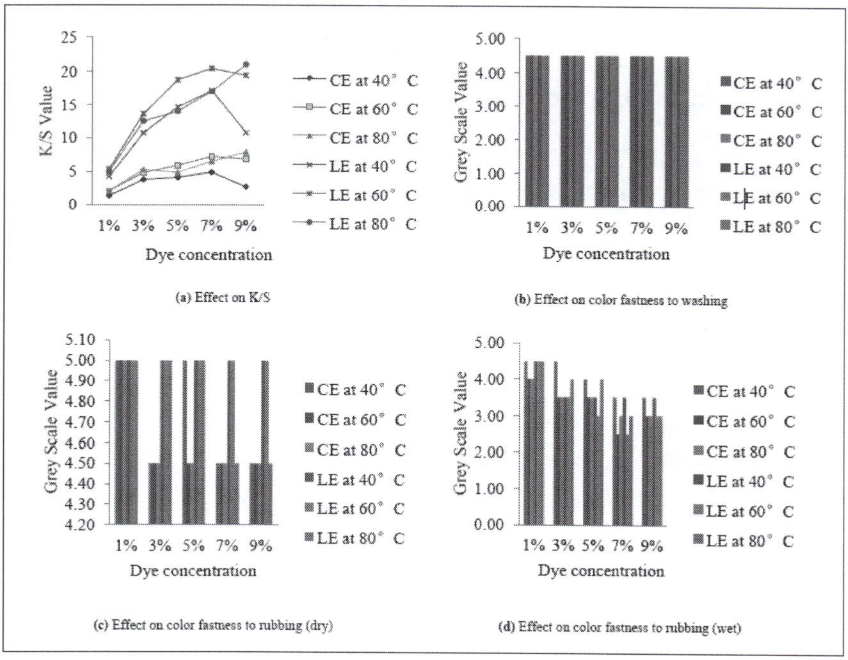

Figure 1. Effect of dye concentration and temperature at 20 g/l salt & 5g/l alkali

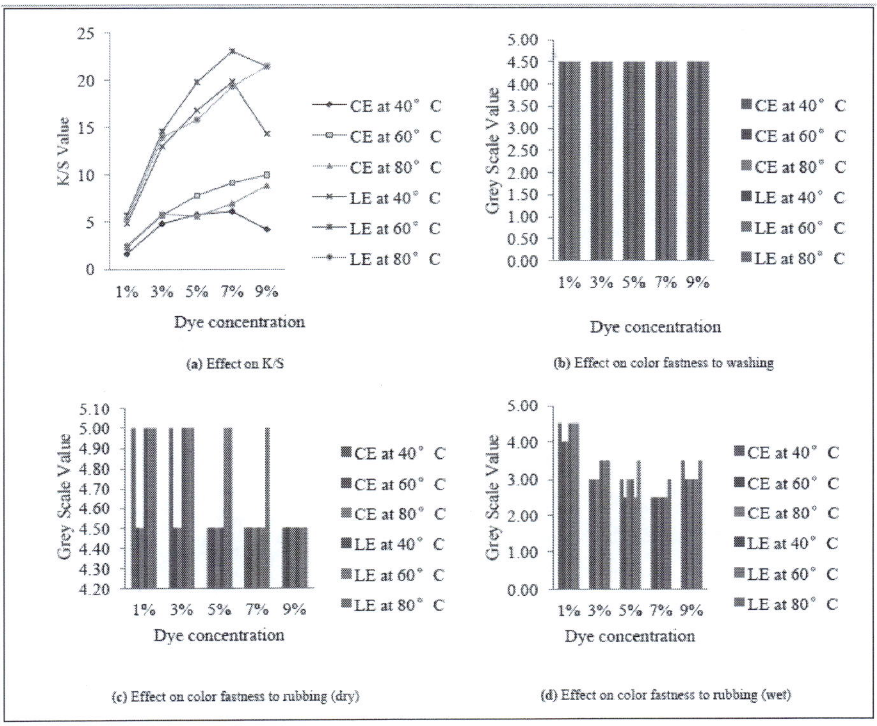

Figure 2. Effect of dye concentration and temperature at 30 g/l salt & 8g/l alkali on color strength & fastness properties

3.1.2. Effect of Temperature on Color Fastness

As the temperature increases, the depth of shade also increases. As a result, when the dyed samples are rubbed against an object, more dye molecule comes out from the fabric surface hence the value become less which means rubbing fastness property is not good. From the Figure 1 (c & d) and Figure 2 (c & d) for both fabrics, the rubbing fastness values are higher at a particular temperature.

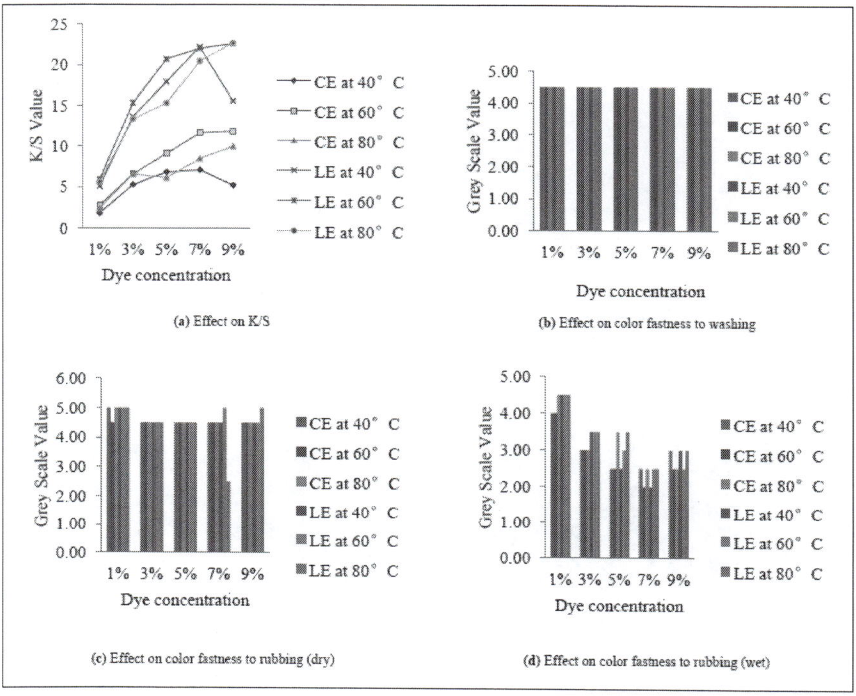

Figure 3. Effect of dye concentration and temperature at 40g/l salt & 10g/l alkalion color strength & fastness properties

In some cases, the rubbing fastness value increases with temperature which may be due to good fixation of the dye molecule at that temperature to these particular fiber. However, the effects of temperature on washing fastness properties of the dye molecule are not as visible as in case of rubbing fastness. Figure.3.b reflects that irrespective of dyeing temperature and fabric type, color staining values to multi-fiber (DW) are same at all temperature for both fabrics.

3.2. Effect of Salt and Alkali on Color Strength and Fastness

3.2.1. Effect of Salt and Alkali on Color Strength

The plotted curves of Figure 1(a), Figure 2(a) and Figure 3(a) indicate that with the increased amount of salts and alkali the dye uptake also increased. The plotted data shows a gradually increased curve with linear relation. The highest value of dye uptake was observed 11.85 for CE and 22.6 for LE with the amount of 40g/l salt and 10g/l of salt and alkali concentration respectively. The least value of salt and alkali concentration i.e. 20 g/l and 5 g/l shows the minimum K/S value for CE and LE fabrics which mean that this concentration has less impact at the values of dye uptake for fabrics. However, the effect of salt and alkali concentration on dye uptake is also affected by other factors such as-dye concentration, dyeing temperature and time etc.

3.2.2. Effect of Salt and Alkali on Color Fastness

The plotted curves in Figure 1(d) & Figure 2(d) indicate that with the increased amount of salts and alkali, the dye uptake also increased for both fabrics as a result more dye molecule comes out during wet rubbing. The higher value of grey scale in Figure 2 (c & d) indicates that dyed sample contained less dye molecule at low concentration of salt and alkali. At low concentration of alkali and salt, the grey scale value for dry and wet rubbing reached upto 4.5-5 which means good rubbing fastness. However, salt and alkali concentration has no significant effect on wash fastness properties of the both samples as shown in Figure 3(c).

3.3. Effect of Dye Concentration on Color Strength and Fastness

3.3.1. Effect of Dye Concentration on Color Strength

K/S value in Figure 1(a), Figure 2(a) and Figure 3(a) reflects that dye concentration has an unavoidable impact on dye uptake for both fabrics. Generally as the dye concentration increases, the availability of the dye molecules in the dye bath also increases as a result more dye molecule can attach to the fiber and hence the higher K/S value is obtained. In case of CE fabric, the results show that K/S value of dyed fabric increased gradually as the dye concentration shifted from 1 to 9% along with others parameters i.e. temperature, salt and alkali concentration with some exceptions. Certainly, the maximum K/S values (11.85) for CE and 22.68 for LE fabric were achieved at 9% dye concentration with some deviations in temperature. The deviation in K/S value at dye concentration 7% and 9% for both fabrics is similar than other values.

3.3.2. Effect of Dye Concentration on Color Fastness

The effect of dye concentration on rubbing fastness properties of the dyed samples for both fabrics is more significant than other parameters. Generally as the dye concentration increases, the availability of the dye molecules in the dye bath also increases as a result more dye molecule can attach to the fiber and hence depth of shade increases. From Figure 1(c & d) and Figure 2(c & d), the result shows that as the dye percentage shift from 1 to 9%, the grey scale value declines because dye molecule from dyed sample easily came out during rubbing. On the other hand, irrespective of dye concentration for both fabrics in Figure 3(b), it can say that color staining values are less. This may be happen due to two factors; for low dye concentration few dye molecules came out during washing as a result staining of multi-fiber was less and for higher dye concentration, dye molecules may be fixed strongly which increased the resistance to washing of the dye molecules.

4. CONCLUSIONS

A correlative study was made on color fastness properties &color strength of cotton-elastane (CE) and lyocell-elastane (LE) for different dyeing parameters. The color fastness values for both fabrics show maximum value i.e.4-5 to 5 at a particular temperature with some exceptions. The fastness values of lyocell-elastane fabrics is more in comparison with cotton-elastane fabrics for dry and wet rubbing which may be due to its high fixation rate. In case of washing fastness properties both fabrics show any differences in their values of 4-5. Therefore, it can be stated that rubbing fastness properties is inversely related with dye concentration excluding washing fastness with respective salt and alkali concentration, dyeing time and temperature. In case of color strength assessment, results showed that highest K/S value for CE fabric was obtained at dye concentration 9%, salt concentration 40g/l, alkali concentration 10 g/l and dyeing temperature 60°C whereas highest K/S value was recorded at dyeing temperature of 80°C with same concentration of salt & alkali in LE fabric. In case of CE fabric, the maximum K/S value has reached up to 11.85 and for LE fabric, K/S was 23.03.

ACKNOWLEDGEMENTS

The authors are thankful to the management of Alim Knit (BD) Ltd for providing the laboratory facilities for this study.

REFERENCES

1. Albercht. W at al. "Lyocellfibres", Chem Fibers Int, 47, (1997): 289-304.

2. Krassig, H., and Structure Cellulose. "Accessibility and Reactivity Gordon and Breach Science Publishers." (1992).

3. Stana-Kleinschek, K., Ribitsch, V., Kreže, T., Sfiligoj-Smole, M., & Peršin, Z. Correlation of regenerated cellulose fibres morphology and surface free energy components. Lenzinger Berichte, 82(1), (2003):83-95.

4. Chavan, R. B., and A. K. Patra. "Development and processing of lyocell." Indian journal of fiber and textile research" 29.4 (2004): 483.

5. Lenz, J., J. Schurz, and E. Wrentschur. "Comparative characterization of solvent spun cellulose and high wet modulus viscose fibers by their long periods." Actapolymerica 43.6 (1992): 307-312.

6. Rattee, I. D. "Reactive dyes for cellulose 1953–1983." Review of Progress in Coloration and Related Topics14.1 (1984): 50-57.

7. Allegre, C., et al. "Treatment and reuse of reactive dye effluents." Journal of Membrane Science 269.1 (2006): 15-34.

8. Haji, M. A.Properties of cotton/spandex single jersey fabrics. The Indian Textile Journal (2013).

9. Taylor, James M., Michael J. Bradbury, and Stephen Moorhouse. "Applied Technology-Dyeing Tencel and Tencel A100 with poly-functional reactive dyes." AATCC Review-American Association of Textile Chemists and Colorists 1.10 (2001): 21-24.

10. Jiangang, L. Dyeing of Spandex/Superfine Polyester Knitgoods. Dyeing and Finishing, 7, (2003):006.

11. Hong_fei, Q. I. A. N., & Xin_yuan, S. O. N. G.. "The Effect of Auxiliaries on Dyeing Spandex With Disperse Dyes. Textile Auxiliaries", 2, (2001): 007.

12. GAO, X. H., & SONG, X. Y. Research on the dyeing and finishing processes of the new cotton/spandex elastic fabrics. Textile Auxiliaries, 6, (2005): 010.

13. Kabir, S. F., Iqbal, M. I., Sikdar, P. P., Rahman, M. M., & Akhter, S. "Optimization of Parameters of Cotton Fabric Whiteness". European Scientific Journal, 10(36) (2014).

14. Shaohua, L. Pretreatment Parameters and Dyestuffs Optimization before Ink Jet Printing. Dyeing and Finishing, 12, (2001):001.

15. QIAN, Q. F., ZHANG, J. F., ZHANG, J. G., & JIN, X. H. Wet processing of polyester/spandex fabric. Dyeing & Finishing, 7, (2013):005.

16. LI, X. J., LI, Y. Y., & MA, D. F. Technology Practice of Dyeing and Finishing Cotton/Spandex Undershirt Cloth. Shandong Textile Science & Technology, 2, (2004):006.

5
CHAPTER

Effects of Metal Filament's Alignment on Tensile and Electrical Properties of Conductive Hybrid Cover Yarns

Ali Asghar[1,2], Mohd Rozi Ahmad[1] and Mohamad Faizul Yahya[1]

[1]*Faculty of Applied Sciences, Universiti Teknologi MARA, Shah Alam, Selangor, Malaysia.*

[2] *Baluchistan University of IT, Engineering and Management Sciences, Quetta, Pakistan.*

ABSTRACT

Of late, a significant amount of research has been carried out using metals for conductive fabrics and composites to obtain the desired level of electromagnetic shielding along with physical and mechanical properties for its durability. However, incorporating these metallic filaments as core or as an integral part of core deteriorate the mechanical properties of textile yarns and ultimately the fabrics'. Moreover, the transparency of fabrics to high frequency/smaller wavelength waves further increases with the attempt to increase the metallic filament diameter. This study, therefore analyses the effects of metallic filament alignment in order to improve the mechanical properties of the resultant hybrid yarns, and devises an alternative method to increase the amount of conductive filament without increasing the diameter of the conducting wire in hybrid yarn structures. The results suggest that the tensile properties of the proposed hybrid cover yarns with conductive filament as covering component is superior as compared to the yarns having a conductive filament in the core, however, the electrical resistance increases with an increase in conductive filament length. The tenacity, elongation and initial modulus values were enhanced several times by incorporating the conductive filament as spiral covering. Moreover, with the proposed alignment method, the amount of copper in hybrid yarn can be increased up to some extent by changing the number of turns of covering component.

Keywords: Copper Conductive yarns Electrical resistance Hybrid cover yarns Metallic filament

INTRODUCTION

With the advent in technology, the reduction in the size of circuits has created smaller but more powerful sources for electromagnetic (EM) radiation emissions. These electronic/electrical devices have now surrounded our everyday environment in close proximity. The radiated emissions are not just responsible for electromagnetic interference (EMI) in between the devices but also have negative influences on human health (Jagatheesan et al. 2014). The electromagnetic radiation is already categorized as the fourth most serious source of public pollution following the noise, air and water pollution (Ozen et al. 2012). This pollution is mainly attributed to the ever-increasing use of wireless communications all around the globe. With the further advancement of information technology, and in order to cope up with the needs, the useable range of EM wave's frequencies will shift to much higher bandwidths and thus will be more problematic as much higher levels of this pollutant will spread in the atmosphere. In order to cope up with the higher frequencies, significant amount of research has been carried out using metals to conductive fabrics and composites to obtain the desired levels of shielding as well as physical and mechanical properties for its durability (Jagatheesan et al. 2014).

Conductive yarns, as per their name sounds, are capable of conducting electricity through them. Usually the textile materials (natural and synthetic), at standard atmospheric conditions, are very good electrical insulators with the exception of only few intrinsically conductive polymers such as polyacetylene, polyaniline, polypyrrole etc., which are capable of conducting charges (Maity et al. 2013). Non-conducting textile materials can be converted to conductive materials by either giving them surface treatments like conductive paints, ionic/electroless plating, vacuum metallization or by incorporating conductive fillers inside in a particular order (Cheng 2006). According to Duran and Kado lu (2014), certain properties of conductive textiles that includes: weight, flexibility, pliability, air permeability, porosity, comfort in wearing, less corrosiveness and lower cost of production, have made its use imperative in applications where electromagnetic shielding and static dissipation is required.

There are certain limitations to every method for the production of conductive fabrics; the limiting factors for intrinsically conductive polymers are poor long-term stability, washing fastness and rigid characteristics (Šaravanja et al. 2015). Similarly, coating of carbon fibres onto a material for conductivity is reported to have very high cost of production, whereas, the use of Carbon Nano tubes is restricted for its uniformity and dispersibility, specially for the Multi-walled Carbon Nano tubes (Jagatheesan et al. 2014). Although it has its own demerits, but producing fabric by incorporating conductive fillers (hybrid yarns) inside is a very popular and common way of producing EM shielding fabrics.

These conductive fillers or hybrid yarns can be formed by using continues filaments like stainless steel or copper wires placed alongside (plied) with non-conducting textile materials, placed as core material with covering of textile materials or in the form of discontinuous cut staple fibres (limiting factor), blended together with some non-conductive natural or synthetic textile materials. Rajendrakumar and Thilagavathi (2012), plied polyester filament with 0.04 mm copper wire on a ring doubling machine to investigate the effects of fabric weave and pick density on electromagnetic shielding effectiveness (EMSE). They concluded that the plain weave posed better EMSE as more number of interlacements reduces the contact resistance, they further concluded that the pick density is directly related to increased EMSE.

Duran and Kado lu (2014), inserted silver coated Polyamide filaments in core via a special arrangement after the front roll delivery system in Ring spinning. They used this hybrid yarn structure in the weft with varying weft densities to analyse its impacts on EMSE, moreover, they investigated different proportions of Ag/PA and concluded that the higher the % age of conductive material in blend the higher will be the EMSE. Several other researchers used somewhat similar techniques to incorporate the conductive filament in core, like, Yu et al. (2014), inserted stainless steel wire of 50 μm in core by using hollow spinning machine, the covering components were Antibacterial Nylon and Bamboo Charcoal Polyester filament. Siro core spinning system with double roving as outer layers and stainless steel core was examined for the effects of orientation of fibres within the structure by Ortlek et al. (2012). Recently, Yu et al. (2015), incorporated stainless steel wire in core using hollow spindle spinning in order to investigate the functional and EMSE of elastic warp knitted fabrics. They concluded that multi-layered fabrics with laminating angles 0°/90° produced better results as compared to 0°/0° layered fabrics due to the formation of metal grids in the arrangement.

Liu and Rong (2015), investigated the effects of metal fibre contents by incorporating a blended yarn having 25 % stainless steel fibres into fabrics with varying warp and weft densities. They concluded that the greater the metal fibre content of a fabric more would be the conductivity and better EMSE.

In general, metal filament/wires have rigid characteristics when compared to textile yarns and filaments. They are resilient and have less efficiency to return to their normal shapes after experiencing mechanical deformations. Moreover, when placed inside the yarn as core component (in case of core-sheath or plied yarns), the mechanical/tensile properties of these metallic filaments drape over the properties of the textile component (Vasile et al.2011). Since, the core component's mechanical/tensile properties in any hybrid yarn structure dominates the yarn's mechanical/tensile properties (Pl et al. 2015), therefore incorporating the metallic filaments as core deteriorate the mechanical/tensile properties of textile yarns and ultimately the fabrics'.

Furthermore, the rigidity of these metallic wires, when inserted as core component, causes openness in fabrics as they are not properly interlaced, hence making the fabric transparent for the smaller wavelength waves (Cheng 2006). The transparency of fabrics to high frequency/smaller wavelength waves further increases with the attempt to increase the metallic wire's diameter, in order to increase the amount of metal inside the fabrics for greater EM shielding effectiveness (Perumalraj and Dasaradan, 2010).

These declining mechanical/tensile properties due to the insertion of metallic filaments in the hybrid yarn's core have been neglected by the researchers so far. Their studies were mainly concentrated around improving the electrical conductance/EM shielding. Moreover, the increase in diameter of the metallic core to increase the EMSE further aggravates the problem. This study therefore analyses the effects of metallic filament's alignment in order to improve the mechanical properties of the resultant hybrid yarns and devises an alternative method to increase the amount of conductive filament without increasing the diameter of the conductive wire in hybrid yarn structures.

METHODS

Materials

Two different genres of copper with respect to their diameters that includes 0.1 and 0.2 mm with Tex counts of 75 and 320.80 Tex respectively, were used as conductive components for the study. The copper wires were obtained from the open market manufactured by FE Magnet Wire Company, Malaysia. The electrical resistance was recorded as 2300 Ω/KM for 0.1 mm diameter and 687 Ω/KM for 0.2 mm diameter. For the counter non-conductive textile part, 31.60 Tex Polyester yarn and 78.52 Tex Acrylic/Wool (AW) yarn (blending ratio of 70/30) were used. The selection of non-conductive textile component such as Polyester and Acrylic/Wool was carried out just to highlight the impacts of alignment of components regardless of their fibre type. Winding of 0.2 mm copper on to a core require the core to be thick or bulky enough for the wire to wound properly. Coarse count AW yarn was selected as it provides the required bulkiness.

Preparation of yarn samples

The two component hybrid cover yarns were produced using a hollow spindle-spinning machine YCHN-303. The two components include a core component and a cover component (Figs. 1, 2). The core component was drawn through the machine's drafting arrangement (without being drafted) from within the hollow spindle to the output roller, whereas, the covering component was wound on hollow spindle's bobbin. Different yarn samples were produced by varying the genre and alignment of conductive and non-conductive components and are represented in Tables 1 and 2. The machine parameters were carefully monitored and controlled as it has considerable influence on the output product. These parameters include the hollow spindle speed, tension of core component and machine throughput for controlling and varying the number of turns of covering.

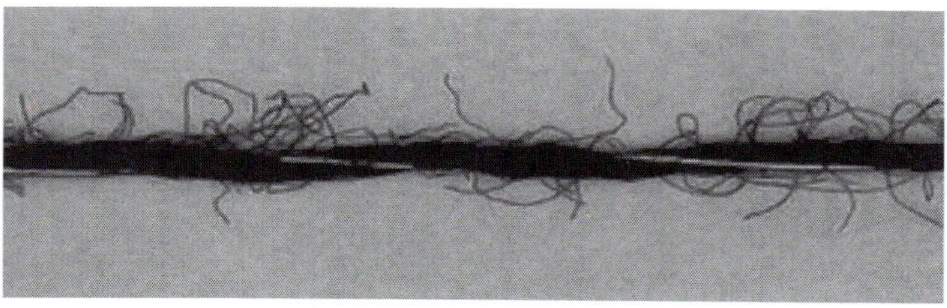

Figure 1. Sample A1 "copper core"

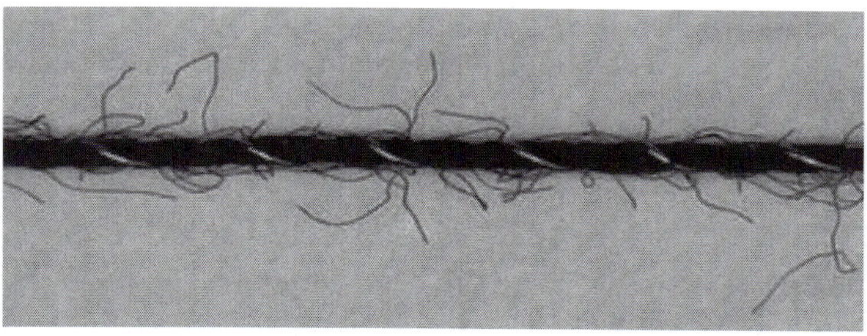

Figure 2. Sample A6 "copper cover"

Table 1. Component definition

Yarn sample ID	Non-conductive textile component	Conductive component
A	31.61 Tex polyester yarn 40/2 cotton count	0.1 mm dia, 75 Tex copper wire
B	78.52 Tex AW yarn (70/30) 28/2 worsted count	0.2 mm dia, 320.80 Tex copper wire

Table 2. Sample classification

Yarn sample number	Core component	Covering component	Turns per meter of covering component (TPM)	Resultant count (Tex)	Amount of core (% age)	Amount of covering (% age)
A1	Copper wire	Polyester yarn	342	89.40	62.34	37.66
A2	Polyester yarn	Copper wire	429	109.45	27.85	72.15
A3	Polyester yarn	Copper wire	504	111.82	27.64	72.36
A4	Polyester yarn	Copper wire	630	112.15	26.85	73.15
A5	Polyester yarn	Copper wire	771	115.50	27.15	72.85
A6	Polyester yarn	Copper wire	941	116.72	26.38	73.62
B1	Copper wire	AW yarn	512	354.56	73.10	26.90
B2	AW yarn	Copper wire	354	415.32	18.27	81.73
B3	AW yarn	Copper wire	433	420.32	18.39	81.61

Tensile testing of yarn

For the tensile properties of conductive hybrid cover yarns, Testometric Tensile tester was utilized following the guidelines of ASTM D2256. According to the standard, the sample breaking time of 20 ± 3 s was achieved at the extension rate of 100 mm/min on the CRE type machine. The load cell used had the maximum loading capacity of 500 N, which is recommended for yarn testing according to the standard. The recommended type of yarn clamping jaws with the gauge setting of 250 mm was used. The yarns were tested and compared to their basic tensile properties, which includes tensile strength, elongation and initial modulus.

Electrical resistance testing

AATCC Test Method 84-2011 was used as the standard procedure for testing the electrical resistance of yarns. KEITHLEY 197 Autoranging Microvolt DMM meter was used for analysing the electrical resistance of 100 mm yarn samples at standard test atmospheric conditions. Since copper wires have very less resistance to the flow of electricity, measurements at 200 Ω range and 450 V were conducted using a 4-Terminal configuration (Fig. 3) in order to minimize errors due to the voltage drop across the test leads. The Kelvin test leads into the input HI/LOW and OHMS SENSE HI/LOW were used. The test leads were short-circuited at Relative (REL) mode to reveal 000.000 ohms resistance (Fig. 4). The resistance meter was calibrated as per the procedures laid down by the operations manual.

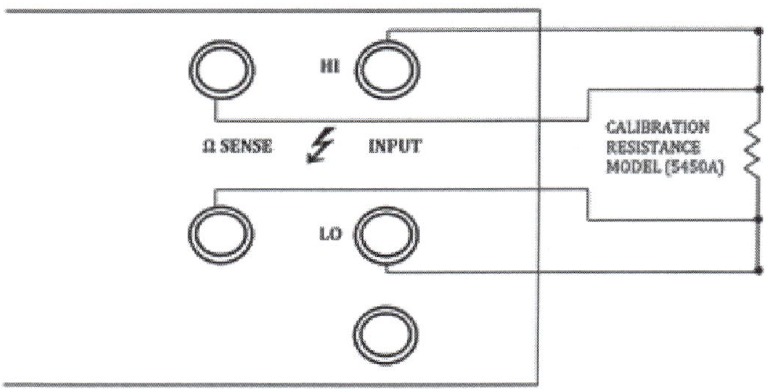

Figure 3. 4-Terminal configuration

Figure. 4. Calibrated multimeter

RESULTS AND DISCUSSIONS

Effects of metal filament's alignment on tensile properties

Tenacity

It is evident from Fig. 5, that the non-conductive textile components have much higher tenacity values as compared to their counter conductive part. Polyester yarn with the tenacity of 31.77 cN/tex was pooled with 2.70 cN/tex copper wire in two different alignments (Figs. 1, 2). The hybrid yarn sample A1, which shows the tenacity of 2.69 cN/tex (Fig. 5), had copper wire as the core component. The spiral polyester yarn on the cover of A1, did not contribute at all to the tenacity of the hybrid yarn structure, in fact, it behaved like a spring that allowed the core to be stretched when axial load was applied. Hence, the sample A1 reflects the tenacity equivalent to the reference copper wire's tenacity.

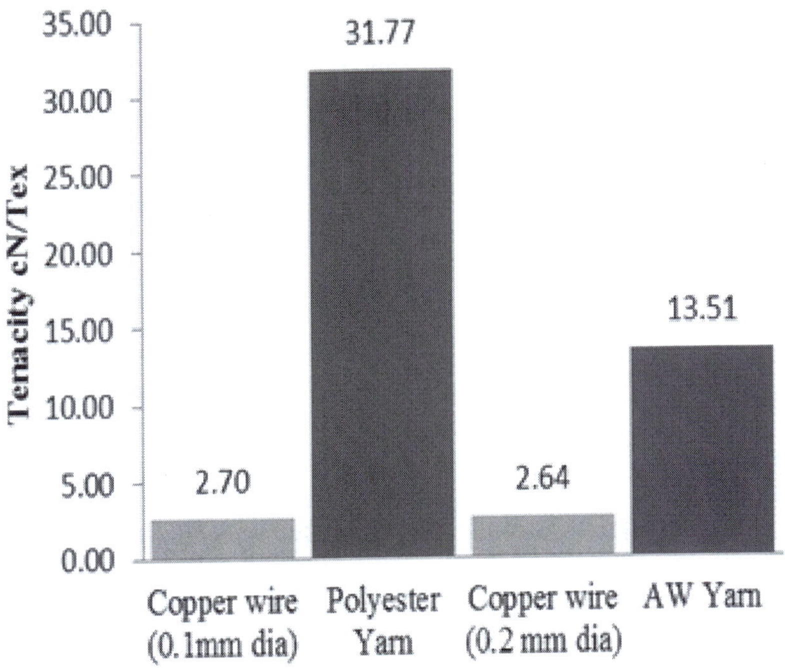

Figure 5. Tenacity of component materials

Sample A2 and A3 have the opposite structural contents as of sample A1, that is, polyester core and copper (Fig. 6) covering with varying twists. The tenacity values for the hybrid yarn samples A2 and A3 are 11.01 and 10.33 cN/tex respectively, which is around four times higher than the tenacity of sample A1.

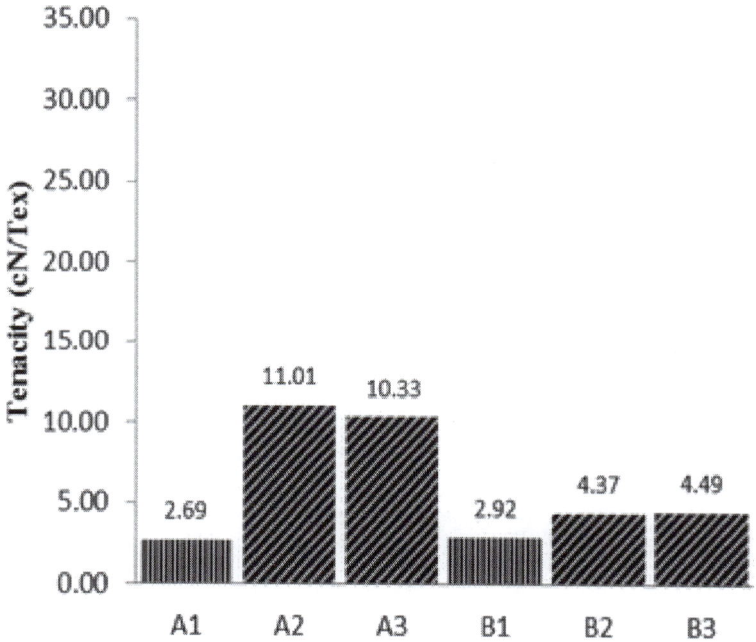

Figure 6. Hybrid yarn's tensile strength comparison

A somewhat similar pattern was observed for AW and 0.2 mm copper hybrid yarns. The sample B1, that had copper as core, showed the tenacity value of 2.92 cN/tex. The copper wire used for preparing B1 has the tenacity of 2.64 cN/tex, which is transferred to the hybrid yarn sample. Again, the AW covering had no or very little contribution towards the tenacity of this hybrid yarn structure. Whereas, the samples B2 and B3, which had the AW core and copper covering, showed higher tenacity values than that of sample B1.

However, the samples A2, A3 and B2, B3 should have displayed the tenacity values almost equivalent to the tenacity of reference core yarns. The reason for reduced tenacity could be the resistance provided by the multiple coils of stiff copper wires. Generally, the more the number of coils the greater should be the tensile strength of the resultant hybrid yarn, since these coils get further tighten up and exert lateral pressure as the axial forces are applied. Therefore, yarn sample A6 with 941 coils per meter has better tenacity values than the two preceding samples as shown in Fig. 7. The R-squared value of 0.9483 verifies this second-degree polynomial regression.

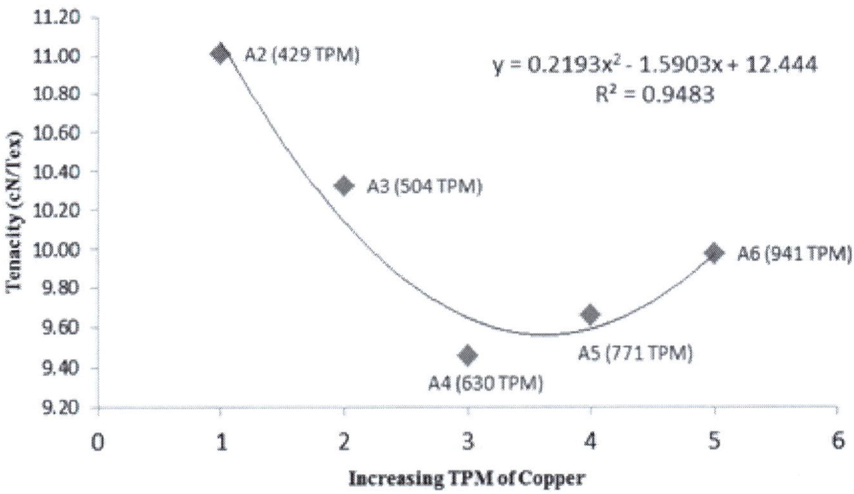

Figure 7. Increasing TPM of copper and tenacity

Elongation

The elongation values of the component materials are displayed in Fig. 8. Both the conductive and non-conductive materials showed somewhat higher elongation values when measured separately. Whereas, when they are combined as hybrid yarns, the samples having copper as a core component (A1 and B1) showed very less elongation (Fig. 9), while the samples having polyester and AW yarn as core components (A2, A3, B2, B3) had very high elongation values (around 4–6 times higher).

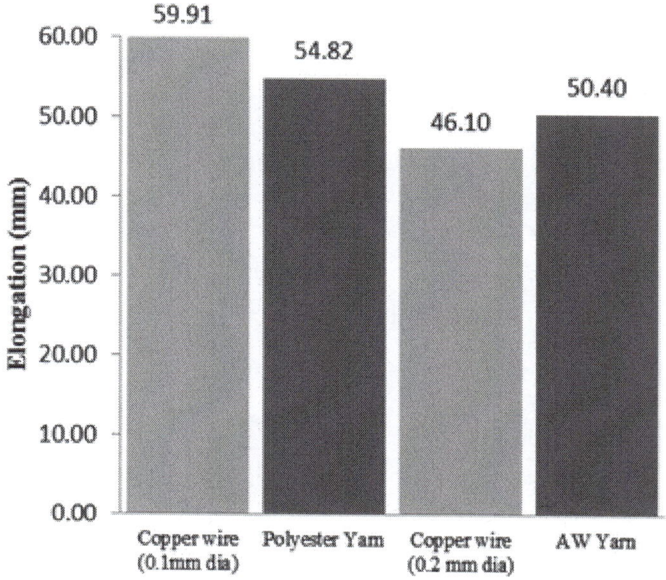

Figure 8. Elongation of component materials

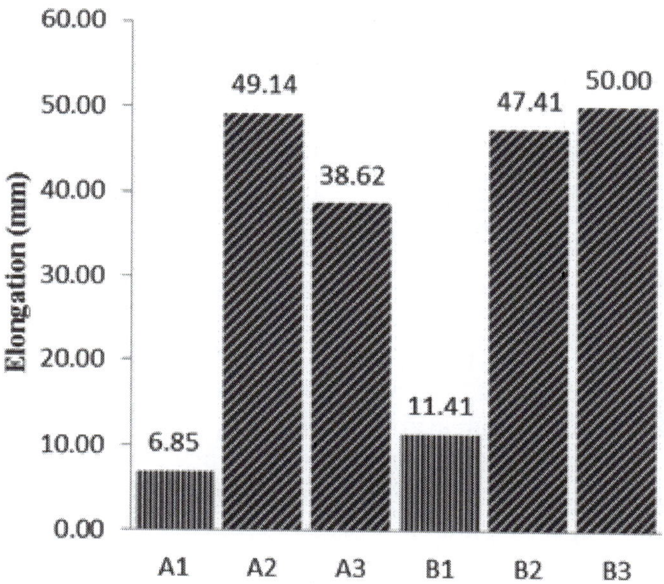

Figure 9. Hybrid yarn's elongation comparison

During hybrid yarn preparation in hollow spinning, the core component had to be kept in high tension for winding the cover properly around the core. To achieve the required tension in case of copper core samples (A1 and B1), a draft of 1.3 between the front drafting roller and the take-up roller had to be maintained. Since copper has very less elasticity due to its high modulus, this draft during spinning made it permanently elongated, hence the elongation values in the hybrid yarn sample A1 and B1 were reduced. This is also reflected in the resultant count of the sample A1 and B1 in Table 2. The resultant count of 75 Tex copper wire and 31.60 Tex polyester turned out to be 89.40 Tex only (sample A1). Similarly, 320.80 Tex copper with 78.52 Tex AW yarn blend in hybrid yarn B1 was only 354.56 Tex.

Initial modulus

The initial modulus values of the component materials are displayed in Fig. 10, where copper showed very high modulus values as compared to the non-conductive textile components. When converted to hybrid yarn structure while keeping copper as a core component (A1 and B1), the initial modulus values of the resultant hybrid yarn remained high, hence making the yarn stiffer and less elastic.

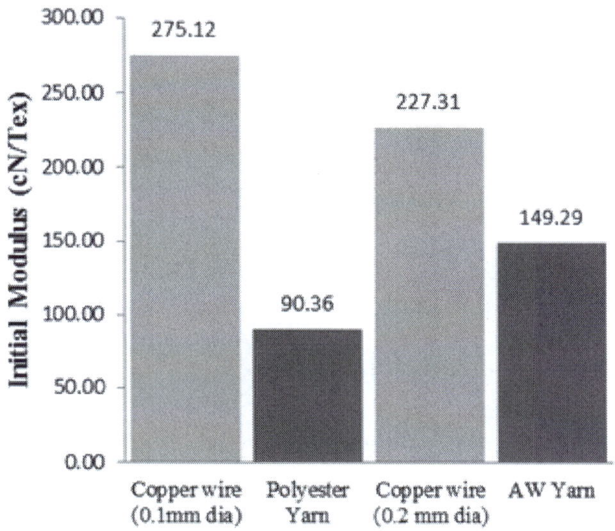

Figure 10. Initial modulus of component materials

Samples A2, A3 and B2, B3 which had copper as covering material and non-conductive textile core, performed otherwise, i.e., showing less modulus values (Fig. 11). Copper as a spiral covering, did not contribute to the modulus values of these hybrid yarns as it can extend and relax itself like that of a helical spring when an axial load was applied. This elastic helical spring made from the coils of copper, even improves the modulus values of the hybrid yarn structure.

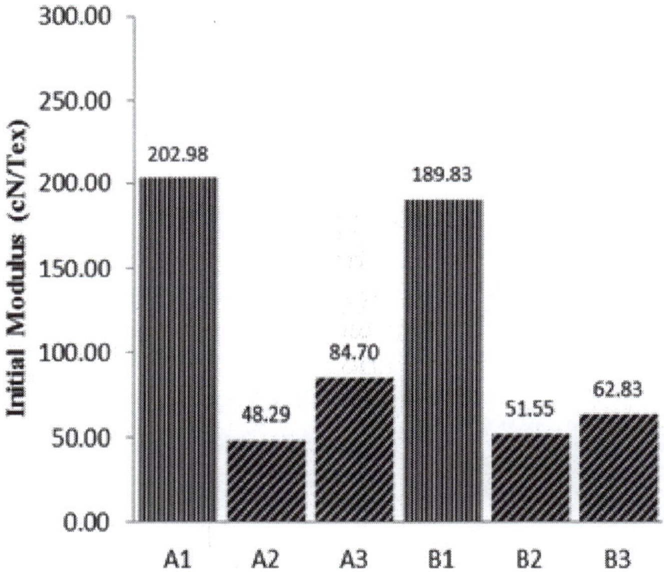

Figure 11. Hybrid yarn's initial modulus comparison

Although there was no significant relation found between turns per meter of copper and initial modulus in this research, however, the number of coils of spring does have a very significant relation to its elasticity, which further needs to be assessed.

Effects of metal filament's alignment on amount of copper

Figure 12 plotted for sample A2 to A6 represents the relation between turns per meter and copper weight in grams per 1000 m.

The R-squared value of 0.9825 for the regression line plotted between the data points clearly shows an increasing linear relation between TPM and increase in copper weight.

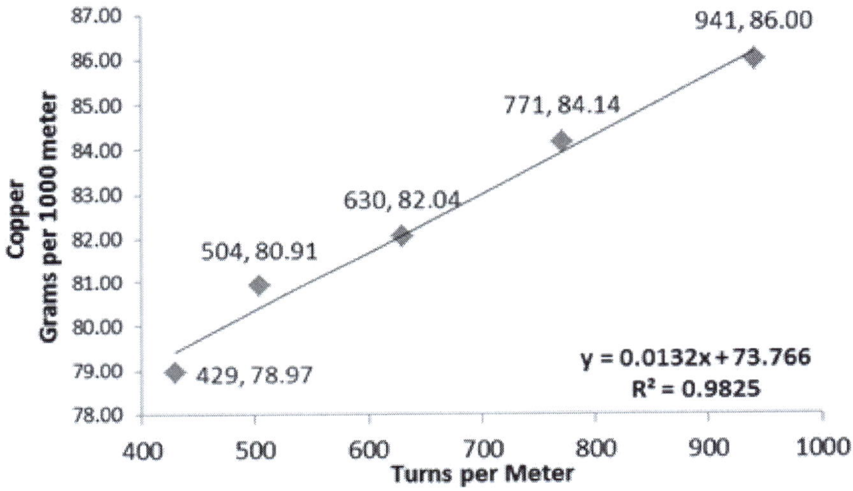

Figure 12. TPM and amount of copper

The values on Y-axis are in weight per unit length, i.e., grams per 1000 m or simply Tex. It can be seen from the figure that a 75 Tex copper can be converted to 86 Tex wire by increasing the number of coils to 941 turns per meter. The resultant Tex count of the hybrid yarn increases as shown in Table 2 (A2–A6), but since the diameter of the resultant yarn remains the same, hence various amounts of copper can be adjusted by changing the number of coils in the hybrid yarn structure for improving the electromagnetic shielding efficiency of the resultant fabrics.

Effects of metal filament's alignment on electrical resistance

Figure 13 represents the relation between the turns per meter and the electrical resistance in ohms per 100 mm of yarn samples at 25 °C and 65 % RH for samples A2 to A6. Although the data points are somewhat scattered along the regression line and the coefficient of determination of 0.8032 also verifies it, still there is an increasing linear relationship amongst the variables. Whereas, the increase in electrical resistance is not remarkable, i.e., only $6.4e^{-3}$ Ω/100 mm increases by increasing the TPM from 429 to 941. The increase in number of coils of copper in the sample A2 to A6 causes increase in length of the copper wire. This can also be verified from the resultant count increase as all the other variables besides the TPM were kept constant. The electrical resistance increases with the increase in length of wire and decreases with the diameter. It can be observed from the Table 3 that the reference copper wire of 0.2 mm showed only 0.0658 Ω/100 mm resistance as compared to 0.23 Ω/100 mm resistance of 0.1 mm copper wire.

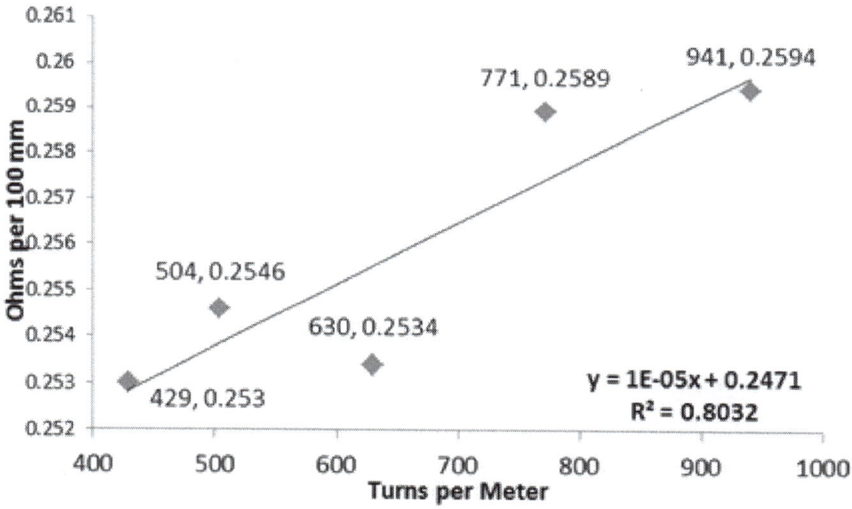

Figure 13 . TPM and electrical resistance of yarn

Sample A1 (copper core) showed a resistance value of 0.2468 Ω/100 mm which is higher than the reference copper resistance of 0.23 Ω/100 mm. Similarly, the sample B1 had 0.0687 Ω/100 mm resistance while 0.2 mm diameter copper wire had 0.0658 Ω/100 mm resistance. This increase of resistance of Sample A1 and B1 is related to the permanent elongation of the copper wire caused by the draft in the delivery zone of the hollow spindle-spinning machine that eventually increased its resistance.

Table 3. Electrical properties of hybrid cover yarns

Sample number	Electrical resistance (ohms per 100 mm)	Amount of copper (grams/1000 m)
Copper wire (0.1 mm dia)	0.2300	75.00
A1	0.2468	55.73
A2	0.2530	78.97
A3	0.2546	80.91
Copper wire (0.2 mm dia)	0.0658	320.80
B1	0.0687	259.20
B2	0.0701	339.44
B3	0.0784	343.10

CONCLUSIONS

The tensile properties of the hybrid cover yarns with conductive filament as covering component is superior as compared to the yarns having a conductive filament in core. The tenacity, elongation and initial modulus values were enhanced several times by incorporating the conductive filament as spiral covering.

Moreover, the amount of copper in hybrid yarn can be altered up to some range by changing the number of turns of covering component. Hence, the disadvantage of incorporating larger diameter wires can be overcome up to certain extent.

The increased coils or TPM of copper has a negative influence as far as electrical resistance is concerned. However, the increase in electrical resistance was not remarkable.

The proposed conductive yarn when converted to fabrics will improve some of the mechanical properties like flexibility, pliability, strength and durability of the conductive fabrics; moreover, the fabric can be woven with higher cover factor to restrict the smaller wavelength waves since the diameter of conductive wires inside the fabrics need not to be enhanced.

REFERENCES

1. Cheng, K. B. (2006). Electromagnetic shielding effectiveness of the twill copper woven fabrics. *Journal of Reinforced Plastics and Composites, 25*(7), 699–709.

2. Duran, D., & Kado lu, H. (2014). Electromagnetic shielding characterization of conductive woven fabrics produced with silver-containing yarns. *Textile Research Journal,*.

3. Jagatheesan, K., Ramasamy, A., Das, A., & Basu, A. (2014). Electromagnetic shielding behaviour of conductive filler composites and conductive fabrics—a review. *Indian Journal of Fibre and Textile Research, 39*(September), 329–342.

4. Liu, Z., & Rong, X. (2015). Influence of metal fibre content of blended electromagnetic shielding fabric on shielding effectiveness considering fabric weave. *Fibers and Textiles in Eastern Europe, 4*(112), 83–87.

5. Maity, S., Singha, K., Debnath, P., & Singha, M. (2013). Textiles in electromagnetic radiation protection. *Journal of Safety Engineering, 2*(2), 11–19.

6. Ortlek, H. G., Alpyildiz, T., & Kilic, G. (2012). Determination of electromagnetic shielding performance of hybrid yarn knitted fabrics with anechoic chamber method. *Textile Research Journal, 83*(1), 90–99.

7. Ozen, M. S., Sancak, E., Beyit, A., Usta, I., & Akalin, M. (2012). Investigation of electromagnetic shielding properties of needle-punched nonwoven fabrics with stainless steel and polyester fiber. *Textile Research Journal, 83*(8), 849–858.

8. Perumalraj, R., & Dasaradan, B. S. (2010). Electromagnetic shielding effectiveness of doubled copper-cotton yarn woven materials. *Fibers and Textiles in Eastern Europe,18*(3), 74–80.

9. Pl, Öİ., Mukavemet, İ. Kİ. N., & Kler, Öİ. (2015). Tensile properties of some technical core spun yarns developed for protective textiles. *Tekstil ve Konfeksiyon (Journal of Textile and Apparel), 25*(2), 104–110.

10. Rajendrakumar, K., & Thilagavathi, G. (2012). Electromagnetic shielding effectiveness of copper/PET composite yarn fabrics. *Indian Journal of Fibre and Textile Research,37*(June), 133–137.

11. Šaravanja, B., Malarić, K., & Pušić, T. (2015). Impact of dry cleaning on the electromagnetic shield characteristics of interlining fabric. *Fibres and Textiles in Eastern Europe, 1*(109), 104–108.

12. Vasile, S., Githaiga, J., & Ciesielska-Wróbel, I. L. (2011). Comparative analysis of the mechanical properties of hybrid yarns with superelastic shape memory alloys (SMA) wires embedded. *Fibres and Textiles in Eastern Europe, 89*(6), 41–46.

13. Yu, Z., He, H., Lu, Y., Zhang, J., Lou, C., Chen, A., & Lin, J. (2015). Functional properties and electromagnetic shielding behaviour of elastic warp-knitted fabrics. *Fibers and Textiles in Eastern Europe, 5*(113), 78–83.

14. Yu, Z.C., Lu, Y.H., He, H.L., Zhang, J.F., Lou, C.W., Chen, A.P., & Lin, J.H. (2014). Antibacterial properties and electrical characteristics of multifunctional metal composite fabrics. *Journal of Industrial Textiles,* (100).

6
CHAPTER

A Comprehensive Analysis on the Efficacy of Antimicrobial Textiles

Mohammad Amir Hamzah[1], Ashraful Islam[2], Md. Ziaur Rahman[3], Md. Hasan Ikbal[1], Md. Eman Talukder[2],Kamrul Hasan[4]

[1]*Department of Textile Engineering, Primeasia University, Dhaka, Bangladesh*

[2]*Department of Textile Engineering, Southeast University, Dhaka, Bangladesh*

[3]*Ministry of Textile & Jute, Textile Vocational Institute, Foridpur, Bangladesh*

[4]*Department of Textile Engineering, City University, Dhaka, Bangladesh*

ABSTRACT

This research paper contains the assessment of the microbial quality of some medical textiles and some export quality fabrics. For the assessment of microbial quality, different samples of the medical textiles and finished fabrics with antimicrobial finish have been collected from different hospitals located in Dhaka city and from different export oriented textile industries in Bangladesh respectively. Microbial quality of the samples have been determined by total viable bacterial count (TVBC), total coliform count (TCC), total fecal coliform count (TFCC), total staphylococcus count (TSC), total Shigella-salmonellae and total fungi count (TFC). The bacterial isolates from different medical textiles and export quality fabrics have been identified on the basis of microscopic appearance of the microorganism. In microbiological analysis of medical textile, four types of bacteria have been found- aerobic heterotrophic, coliform, fecal coliform, staphylococcus aureus and fungi. After testing the antimicrobial finished fabric for potency, it has been found that the fabrics can prevent the growth of bacteria like aerobic heterotrophic, coliform, fecal coliform, staphylococcus aureus and fungi but their inhibition growth rate is not satisfactory.

Keywords: Medical textile, Antimicrobial finished textile, Microbial quality, Microorganism, Antimicrobial potency

1. INTRODUCTION

Bangladesh earns nearly about 80% foreign currency from the textile sectors, exporting casual wears as well as functional clothing. Bangladesh exports the textile products mainly to European region. But, now-a-days this field has become very competitive and facing so much challenges in its business in the competitive global arena. Since most of the raw materials are being imported from abroad and therefore Bangladesh textile sector have been facing two important issues of specifically price and quality of the end products. It is noticeable that the European Buyers are very much conscious about the products they import, particularly the health and hygiene aspects of the end products. Under such situation the producers of textile products in Bangladesh have to be very conscious about the health and hygiene property i.e. antimicrobial property of exportable fabrics. The exportable textiles for medical sectors must be treated with antimicrobial finish and they must have the antimicrobial capability against the activity of the microbes. The term 'antimicrobial' refers to a broad range of technologies that can provide varying degrees of protection for textile products against microorganisms. Antimicrobials are defined as the agents that either kill microorganisms or simply inhibit their growth. Antimicrobial finish is a recent innovation in textile finishes. The consumers are now increasingly aware of the hygiene life style and there is a necessity and expectation for a wide range of textile products finished with antimicrobial properties. This kind of finishes prevents the growth of bacteria and unpleasant odor [1].

The degree of activity of antimicrobial agent is denoted by, Cidal (agent that kill microorganisms and static agents that inhibit microorganisms' growth). Antimicrobial textiles can act in two distinct ways of by contact and by diffusion. Disease causing bacteria comes to human health via microorganisms, while clothes are in close contact with the skin, long periods between washings and a special microclimate can favor a fast growth of germs. They come to the human health, penetrate inside the skin and damage the host tissues. Fungi, algae, and viruses which can not be seen by naked eyes can cause serious diseases to human health. The growth of microbes aggravates particularly in humid and warm environment. The result of microbial attack is the staining and loss of the performance properties of textile substrates [2-3].

Antimicrobial finish is used on textile materials to protect the wearer and the textile substrates by controlling bacteria, fungi, mold, mildew and algae. Antimicrobial treatment for textile materials is necessary to avoid cross-infection by pathogenic microorganisms, to control the infestation by microbes, to save the textile materials from staining, discoloration and deterioration, to control the formation of odors, allergies and other health concerns that they cause. Microorganisms cause problems with textile raw materials and processing chemicals, wet processes in the mills, roll or bulk goods in storage, finished goods in storage and transport and goods as they are used by the consumers. This can be extremely critical to operator a clean room, a medical facility, or a food processing facility or it can be an annoyance and aesthetic problem to the athlete or normal consumers. Antimicrobial agents kill microorganisms or inhibit their growth by Cell wall damage, Inhibition of cell wall synthesis, Alteration of cell wall permeability, Inhibition of synthesis of proteins and nucleic acids, Inhibition of enzyme action. A microorganism is an organism that is microscopic. Most microorganisms are single-celled or unicellular. Microorganisms are responsible for the degradation of cotton. They never attack the substrate directly where they live. They produce very complex non-living molecules called enzyme which can degrade the cellulose structure. Enzyme forms complicated chemical reactions under mild conditions. The steps of degradation may occur including the separation of different structural units and then hydrolytic break down of single chains at glycoside rings and then finally converts into carbon dioxide (CO2) and water while enzymes degrade cellulose chains [2, 4, 6].

1.1. Textile Products Used in Medical Sector

Any textile structure which has been designed and produced for the uses of medical applications, including implantable applications is to be called medical textile. The engineering approach to develop textile products that will be suitable for medical and surgical application should possess a combination of the properties including strength, flexibility, antimicrobial and moisture and air permeability. The current uses of textiles in Bangladesh may broadly be classified into two groups. One is Non-plantable materials and the other is health and hygiene products.

Non-plantable materials are used for external application on the body and may or may not make contact with skin. Some examples of this kind of products are Wound care products, Bandages, Gauze, Lint/Cotton.

Solving the problem of healthcare-associated infections and occupationally acquired infections involves an equation with many complex variables. One of the key components is healthcare workers (HCWs), such as doctors, nurses, laboratory personnel and technical professionals, who are frequently exposed to blood and body fluids. These fluids can transmit bacteria that cause colonization or infection, including multi-drug-resistant organisms (MDROs) such as meticillinresistant Staphylococcus aureus (MRSA), Acinetobacter spp. and Enterobacteriaceae (e.g. Escherichia coli, Klebsiella pneumoniae). There is also a risk of transmission of viruses, including noroviruses, respiratory viruses and bloodborne viruses (human immunodeficiency virus, hepatitis B and C viruses), that can survive for hours or days on surfaces [8-10].

Healthcare and hygiene products are important sectors in the field of medicine and surgery. The range of products available is vast but typically they are used either in the operation theatre or on the hospital ward for the hygiene, care and safety of staff and patients. Textile materials used in the operating theatre include surgeon's gowns, caps and masks, patient drapes and cover cloths of various sizes. It is essential that the environment of the operating theatre is clean and a strict control of infection is maintained.

1.2. Factors Affecting the Growth of Microorganism in Fabric

Storage temperature, pH, Relative Humidity of environment, Moisture content, Laundering are the most important controlling factors in case of fabric storage to avoid the growth of microorganisms. Microorganisms grow over a very wide range of temperatures (250c – 300c). The quantity of the textile product must be taken into account in selecting a storage temperature at refrigerator temperature or below. The relative Humidity (RH) of the storage environment is also an important parameter to maintain. Fabric that undergoes surface spoilage from molds, yeast and certain bacteria should be stored under conditions of low RH. It is now generally accepted that the water requirements of microorganisms should be defined in terms of water activity. Laundering also causes little observable surface damage to cotton fiber but it is observed that damage to cotton fiber by microbial attack increases with each laundering. Severe damage is noticed after 10th laundering, suggesting that possible physical breakdown may make the fibers more accessible to microbial attack [3, 5].

1.3. Modern Applications on Cotton Fabrics for Antimicrobial Property

As the awareness of healthcare increasing and the global infectious disease events frequently happening, a wide range of researches has been done to obtain antibacterial textiles which could not only avoid discoloration and degradation of fabrics by microorganisms but also effectively prevent the propagation of bacteria. The inherent chemical and morphological properties of cellulose fibers are reflected in their biodegradability and high hydrophilicity, providing an excellent medium for bacterial growth. Therefore, cellulose products with antimicrobial properties can only be obtained by appropriate chemical modification of the fibers. Cellulose fibers with antimicrobial properties are an essential textile material that can be used in various types of products with applications in apparel, technical, medical and hygienic purposes as well as in the food industry [11].

2. MATERIALS AND METHODS

2.1. Source of Fabric Collection

Medical Textile: Medical textile samples have been collected from different hospitals, situated in Dhaka city (Samples have been collected by ensuring the authority that their name would not be used anywhere in the paper.)

Export Oriented Fabrics: Reedisha Knit Composite and Square Knit, 36, Shahid Tajuddin Ahmed Sarani, Tejgaon Industrial Area, Dhaka-1208, Bangladesh.

2.2. Description of the Samples

Eight samples of medical textiles (Table 1) and three samples of antimicrobial finished exportable fabrics (Table 2) have been examined for this study.

Table 1. Description of medical textiles samples

Sample no.	Description
S-1	Unused pillow case
S-2	Unused bed cover
S-3	Used pillow case
S-4	Used bed cover
S-5	Unused surgical thread
S-6	Used surgical thread
S-7	Unused cotton
S-8	Unused gauge

Table 2. Description of the export oriented fabric samples

Sample no.	Description
S-1	Knit fabrics (s/j), 160 gsm
S-2	Knit fabrics, (s/j) 160 gsm
S-3	Knit fabrics, (s/j), 160 gsm

2.3. Sample Preparation Procedure

The collected samples inserted in to the conical flask, containing 250 ml of sterilized water. In this way, the required number of samples were taken and transferred to the microbiological laboratory for investigation.

2.4. Test Techniques

Two test techniques have been followed:

- Microbiological analysis of medical textiles.
- Antimicrobial potency of finishing agents used on exportable textile cloths.

2.4.1. Test Procedure for Microbial Analysis of Medical Textiles

Sample preparation:

One gram of the sample was weighed aseptically into a sterile test tube and 9 ml of sterile normal saline was added into it. One gram of homogenate was transferred to a test tube containing 9 ml sterile normal saline to make 10-2 dilution and was shaken well with vortex mixture. A sterile dilution up to 10-3 was also made in the same process.

Sterilization:

All equipment and glassware were sterilized by autoclaving at 121°C temperature and 15 p.s.i. for 15 minutes in the microbiology laboratory. All growth media and solutions were sterilized by the same process. Fresh agar plates were prepared at room temperature, stored at 4°C until use and dried at room temperature under aseptic conditions before inoculation of samples.

Growth media:

Dried media such as nutrient agar, mFC agar, Mac-Conkey agar, potato dextrose agar, mannitol salt agar (MSA) and Shigella-Salmonellae agar (SS-agar) were prepared in the microbiology laboratory.

Maintenance of culture, reagents and solutions:

Preservation and maintenance of different materials including microbial stains, heat labile chemicals and reagents etc. were carried out in a refrigerator at 4°C temperature. Nutrient agar slants were used for the routine maintenance of the pure bacterial strains. The pure bacterial cultures were stored at 4°C until use.

All culture was inoculated under aseptic conditions. The organisms were inoculated in a laminar airflow cabinet. The inoculated solid and liquid media for static growth conditions were incubated using binder indicator. In incubator Petridis is kept overnight at 37°C. Then a colony counter was used to count the microbial colonies on solid agar plates.

2.4.2. Test procedure for Antimicrobial Finishing Agents Used on Exportable Textile Cloths

The fabric sample was cut into small pieces (1 sq. Inch) in aseptic condition and kept them in the sterilized conical flask containing 100 ml of sterilized water. The conical flask was kept at room temperature for 48 hours for the diffusion of antimicrobial agent to the water. The known microbial samples were prepared with normal saline and diluted for the microbial counts. The same dilution was also used for the challenge test. Diffused solution from fabrics was then used as samples for the potency test. 100 ml of microbial sample (known microbial counts) of diffused was then separated on to the agar medium and 100 ml of diffused fabric solution as a challenge was added to find out the antimicrobial potency which had been added in the supplied fabric samples.

2.5. Examining the Presence and Measure of Microorganisms or Bacteria in Different Textile Products

The bacterial count was done by standard method of ICMSF. The microbiological condition of safety and hygiene were examined by using the methods recommended by ICMSF.

2.5.1. Total Viable Bacterial Count (TVBC)

Total Viable Bacterial Count test was carried out to find the total microbial counts and for the quality assessment of the products [5].

Total viable bacteria count was carried out by the spread plate technique. Petri plates were sterilized in hot air oven at 180°C for two hours. All types of media used after sterilization, autoclaved at 121°C with 15 p.s.i. for 15 minutes. The homogenate sample (0.1 ml) of each dilution was taken onto each Petri dish. Then the sample was finely spread on the solid nutrient medium. The prepared dishes were then incubated at 37°C for 24 hours for counting mesophilic bacteria. The plates were then screened (using a colony counter) for the presence of discrete colonies after the specific incubation period and the actual numbers of bacteria was estimated as cfu/gm and finally the results per dilution will be recorded. All colonies on dishes containing 30-300 colonies were counted using a colony counter. The actual number of bacteria was estimated as cfu/gm and the results per dilution counted were recorded. Figure 1 shows the viable bacteria containing plate from which total viable bacteria was counted. Viable bacteria has been detected on plate count agar in all the 8 samples of hospital materials.

2.5.2. Total Coliform Count (TCC)

Total Coliform Count (TCC) test was carried out to find the presence of pathogenic bacteria on textile materials, which are responsible for food poisoning, diarrhea, etc. [5].

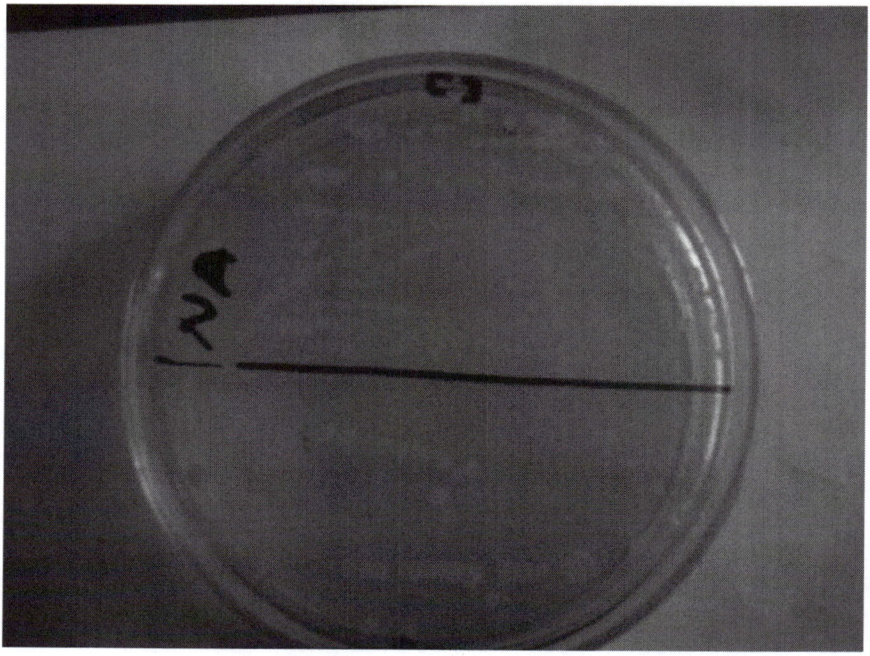

Figure 1. Assessment of Total Viable Bacteria Count (TVBC) on different Medical Textiles

Mac-Conkey agar media was used for counting of total coliform bacteria. Homogenate sample (0.1 ml) from each dilution was taken onto each petri plates. Then the sample was finely spread onto the solid medium and the prepared dishes were incubated at 37°C for 24 hours. Then the plates were screened (using a colony counter) for the presence of discrete colonies after the specific incubation period and the actual numbers of coliform was estimated as cfu/gm. Then the results per dilution were recorded. Figure 2 shows the detection of total coliform containing agar plate from which total coliform was counted. The total coliform count was detected by using Mac-Conkey agar media and it has been found in all 8 tested samples of medical materials.

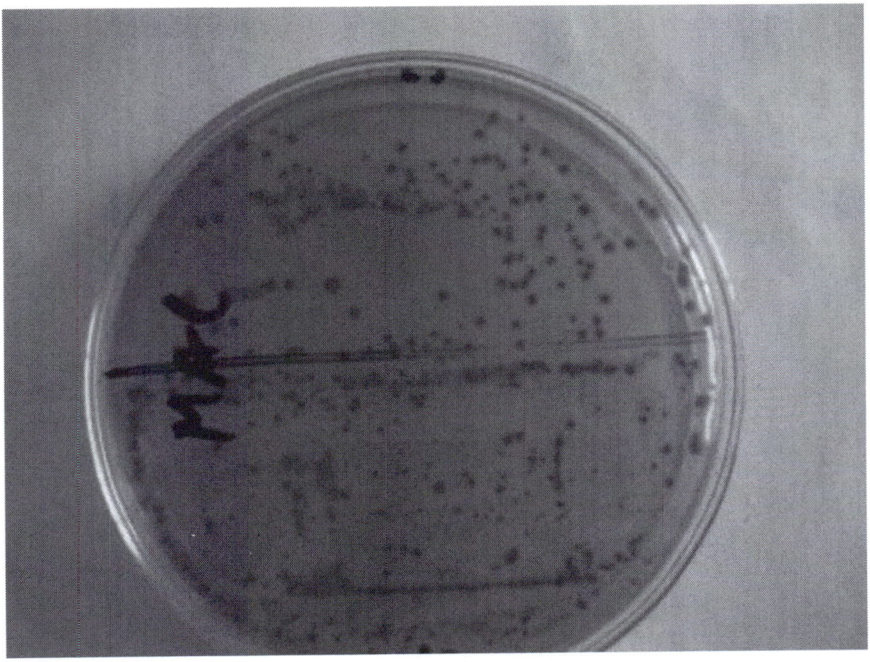

Figure 2. Assessment of Total Coliform Count (TCC) on different Medical Textiles

2.5.3. Total Fecal Coliform Count (TFCC)

Total Fecal Coliform Count (TFCC) test has been done to find out the presence of pathogenic bacteria, which are responsible for diarrhea, bloody dysentery etc. [5].

mFC agar medium was used for counting total fecal coliform. Homogenate sample (0.1 ml) from each dilution was taken onto each sterile Petri dish. Then the sample was homogeneously distributed on the plate using a glass spreader in a backward and forward movement while rotating the plate and the plates were incubated at 45°C for 24 hours. Then the plates were screened for the presence of discrete colonies after the specific incubation period and the actual numbers of Fecal coliform were estimated as cfu/gm. Then the results per dilution were recorded. Figure 3 shows the fecal coliform containing plate from which fecal coliform was counted. No fecal coliform has been found in the samples of medical textile materials.

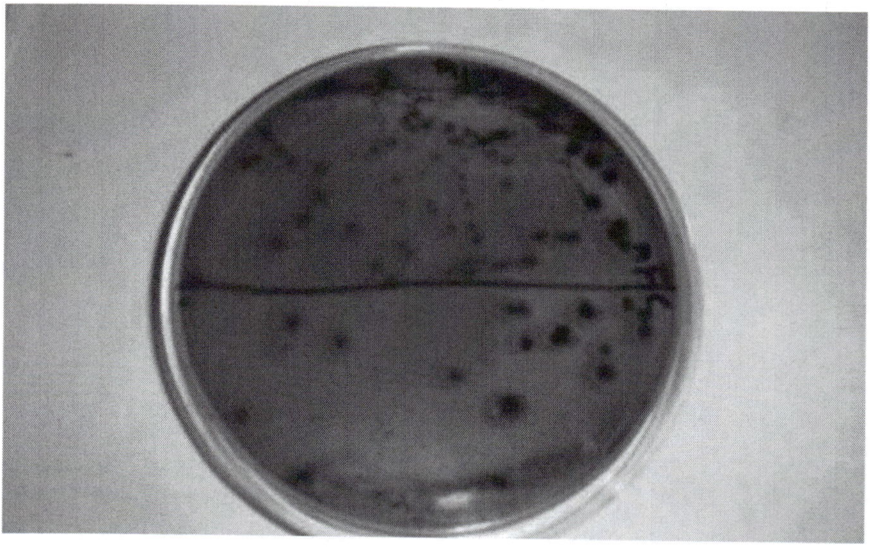

Figure 3. Assessment of Total Fecal Coliform Count (TFCC) on different Medical Textiles

2.5.4. Total Shigella-Salmonellae Count (TSSC)

Total Shigella-Salmonellae Count (TSSC) test has been performed to find out the presence of deadly pathogenic bacteria. The Salmonellae may be transmitted by water, air, food or more really direct anal-oral transmission. Typhoid fever is caused by Salmonellae typhi [5].

Figure 4 shows the Shigella-Salmonellae containing plate from which Shigella-Salmonellae was counted. No Shigella-Salmonellae has been found from the samples of medical textiles.

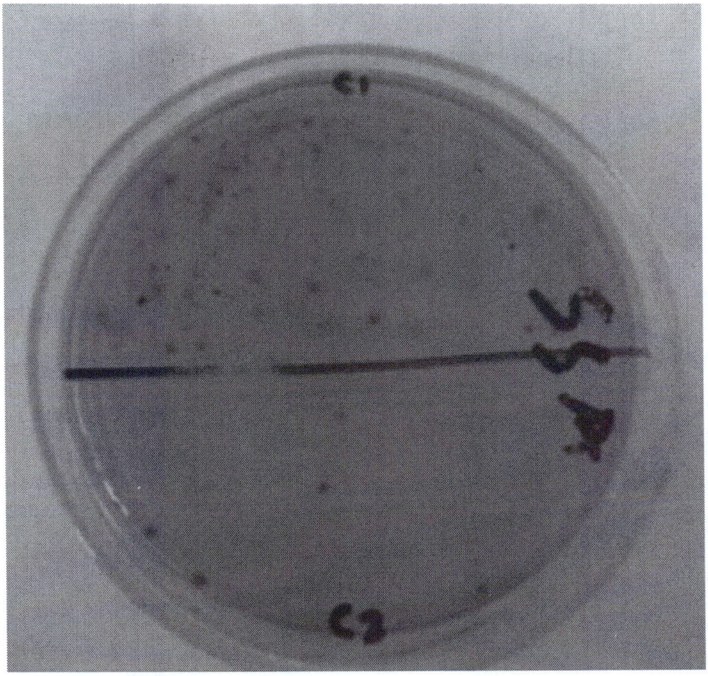

Figure 4. Assessment of Total Shigella-Salmonellae Count (TSSC) on different Medical Textiles

2.5.5. Staphylococcus Aureus Count

Staphylococcus Aureus Count testing was carried out to observe the presence of puss forming bacteria. Levels of Staphylococcus Aureus greater than or equal to 104 cfu/gm are considered as potentially hazardous for skin [5].

Mannitol salt agar medium was used for this purpose. Homogenate sample (0.1 ml) from each dilution was taken onto each sterile petri dish. Then the sample was homogenously distributed on to the plate using a glass spreader in a backward and forward movement while rotating the plate and then the plates were incubated at 37°C for 24 hours and screened for the presence of discrete colonies after the specific incubation period. Figure 5 shows the staphylococcus containing plate from which Staphylococcus aureus was counted. By using Mannitol salt agar medium staphylococcus aureus count of different hospital materials were detected.

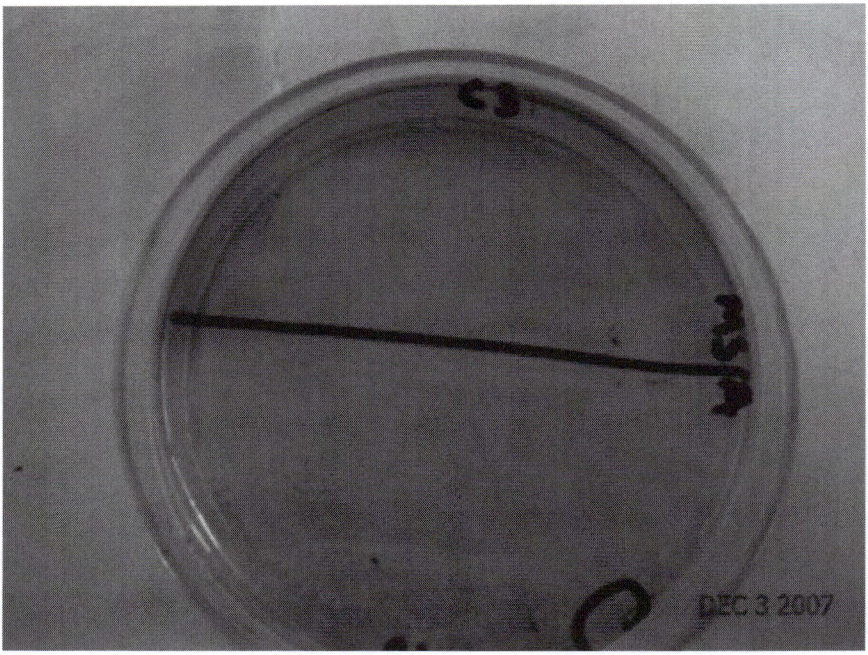

Figure 5. Assessment of Staphylococcus Aureus Count on different Medical Textiles

2.5.6. Total Fungi Count (TFC)

Total Fungi Count (TFC) test has been done to observe the presence of yeast and mold that are responsible for skin disease [5]. By using potato dextrose agar medium, fungi count was detected from different hospital materials. Sabouraud dextrose agar medium was used for this purpose. Homogenate sample (0.1 ml) from each dilution was taken onto each sterile petri dish. Then the sample was homogenously distributed on the plates using a glass spreader in a backward and forward movement while rotating the plate. Then the plates were incubated at 25°C for 5 days. Then all the colonies on dishes were counted using a colony counter and the actual number of fungi was estimated as cfu/gm. The results per dilution were recorded.

3. RESULT AND DISCUSSION

3.1. Microbiological Test Result of Medical Textiles

Table 3. Microbiological analysis of medical textiles

Sample No	Total (TVBC) counts/gram	Total (TCC) coliform/gram	Total Fecal coliform (TFCC)	Shigella - Salmonellae (TSSC)	Total Staphylococcus aureus count	Fungi (TFC)
S-1	1.2 × 104	-	-	-	2.0 × 103	2.0 × 103
S-2	1.8 × 104	-	-	-	6.0 × 103	1.8 × 103
S-3	1.0 × 104	1.8 × 103	-	-	8.0 × 103	6.0 × 103
S-4	1.2 × 104	2.4 × 103	-	-	9.0 × 103	6.4 × 103
S-5	5.0 × 104	-	-	-	3.5 × 103	2.0 × 103
S-6	2.0 × 105	-	-	-	2.2 × 103	1.0 × 103
S-7	9.4 × 104	-	-	-	1.0 × 103	1.0 × 103
S-8	6.8 × 104	-	-	-	5.0 × 103	3.0 × 103

3.2. Potency of Antimicrobial Agents Used on Fabrics

Table 4. Potency of Antimicrobial Agents used on Fabrics

Potency % of antimicrobial agent against different bacterial						
Sample No	Aerobic heterotrophic bacteria	Total Coliform	Fecal Coliform	Shigella - Salmonellae	Staphylococcus aureus	Fungi
S-1	38%	22%	16%	14%	25%	30%
S-2	33%	18%	14%	12%	22%	35%
S-3	28%	22%	18%	18%	26%	28%
Desired	90%	90%	90%		90%	

3.3. Microbial Result Analysis of Medical Textiles

3.3.1. Total Viable Bacteria Count

From Figure 6 it can be seen that, the minimum count was 1.0 × 104 and the highest count was 2.0 × 105. The standard tolerance for Total Viable Bacteria Count (TVBC) should be 5.0 × 104.

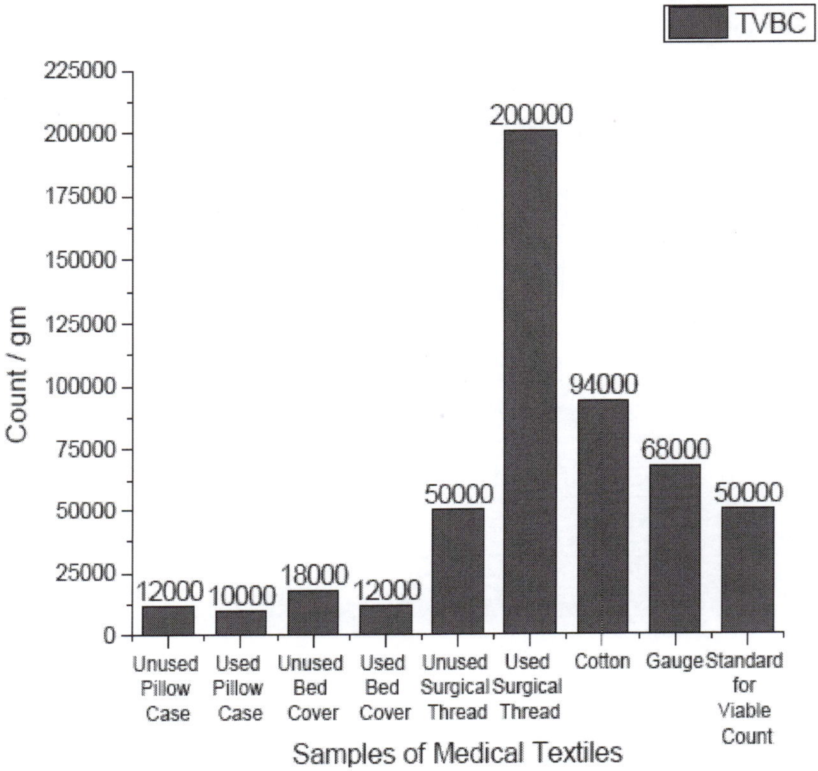

Figure 6. Total Viable Bacteria Count on Medical Textile

3.3.2. Total Coliform Count

From Figure 7 it can be seen that, total coliform were found in 2 samples only-used pillow case and used bed cover. The standard value for Total Coliform Count (TCC) should be only 3.

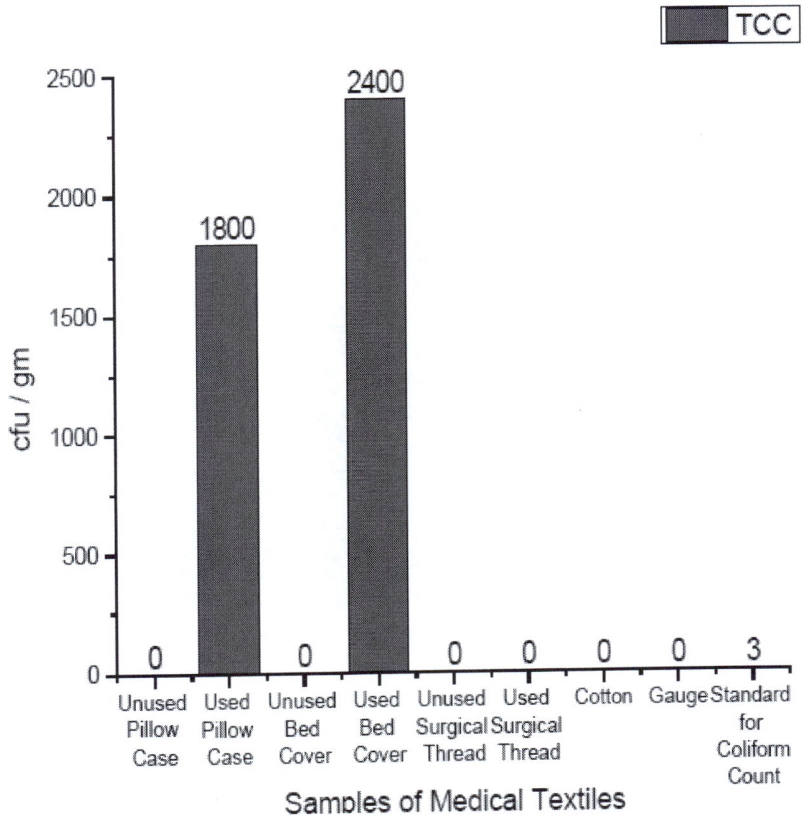

Figure 7. Assessment of Total Coliform count on Medical Textile

3.3.3. Staphylococcus Aureus Count

From Figure 8 it can be seen that, the total number of staphylococcus aureus in 8 samples of hospital materials was detected on plate count agar. Staphylococcus aureus were found in all the samples. The minimum count was 2.0 × 103 and the highest count was 3.5 × 104. The standard tolerance for Staphylococcus aureus should be 0.

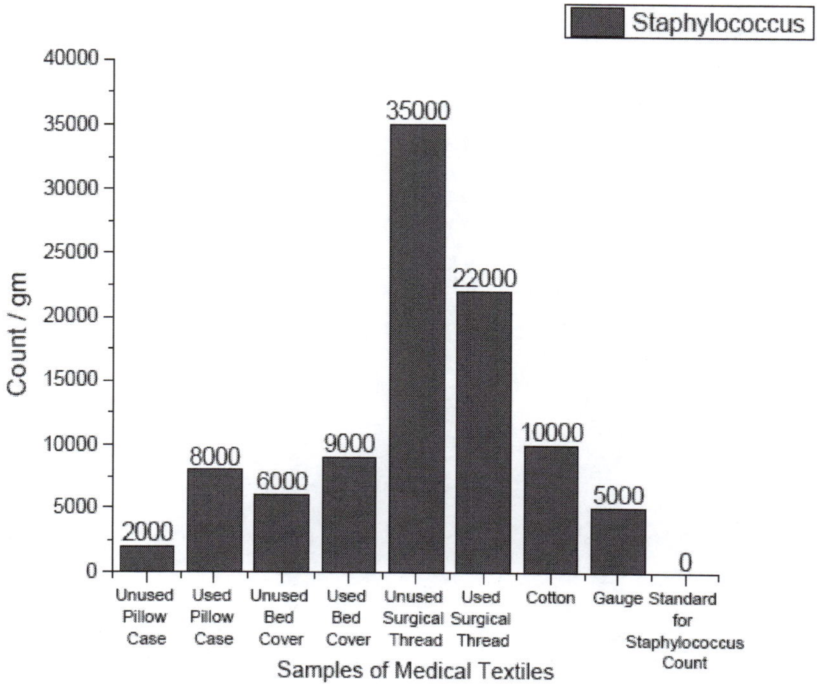

Figure 8. Assessment of Staphylococcus Aureus Count of medical textile

3.3.4. Fungi Count

From Figure 9 it can be seen that,. the total number of fungi in 8 samples of hospital textile materials was detected on count plate agar. Fungi were found in all the samples. The minimum count was 1.1 × 103 and the highest count was 6.4 × 103. The standard value for fungi count should be 0.

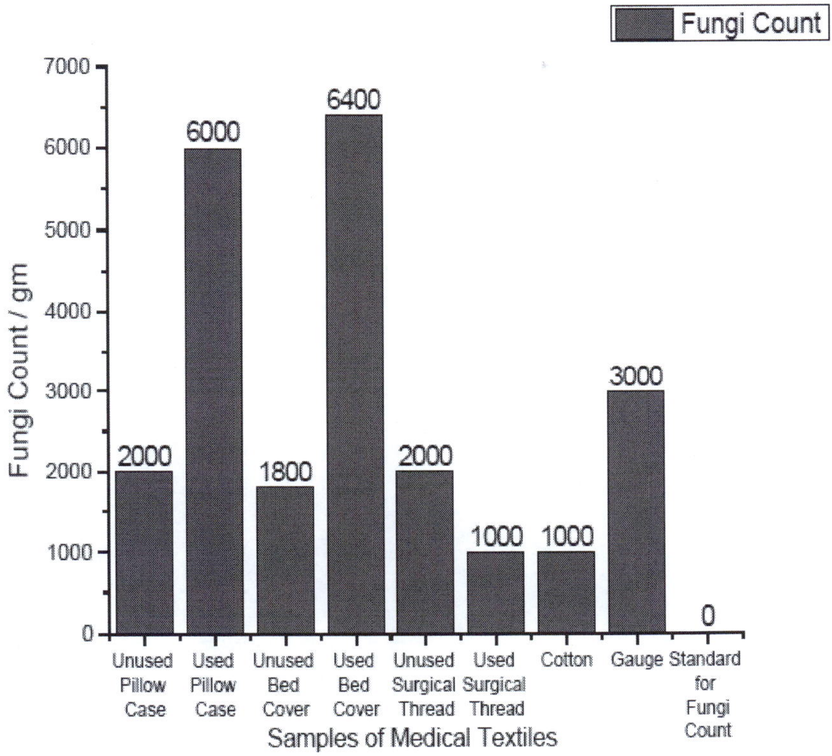

Figure 9. Assessment of Staphylococcus Fungi Count of medical textile

3.4. Result Analysis (Potency of Antimicrobial Agents Used in Fabrics)

3.4.1. Aerobic Heterotrophic Bacteria

From Figure 10 it can be seen that, sample 1, 2, & 3 can prevent 38%, 33% and 28% Aerobic heterotrophic bacteria respectively. The more the sample can inhibit the growth of bacteria means antimicrobial finish is successful. The desired rate of inhibition the growth of bacteria should be 90%.

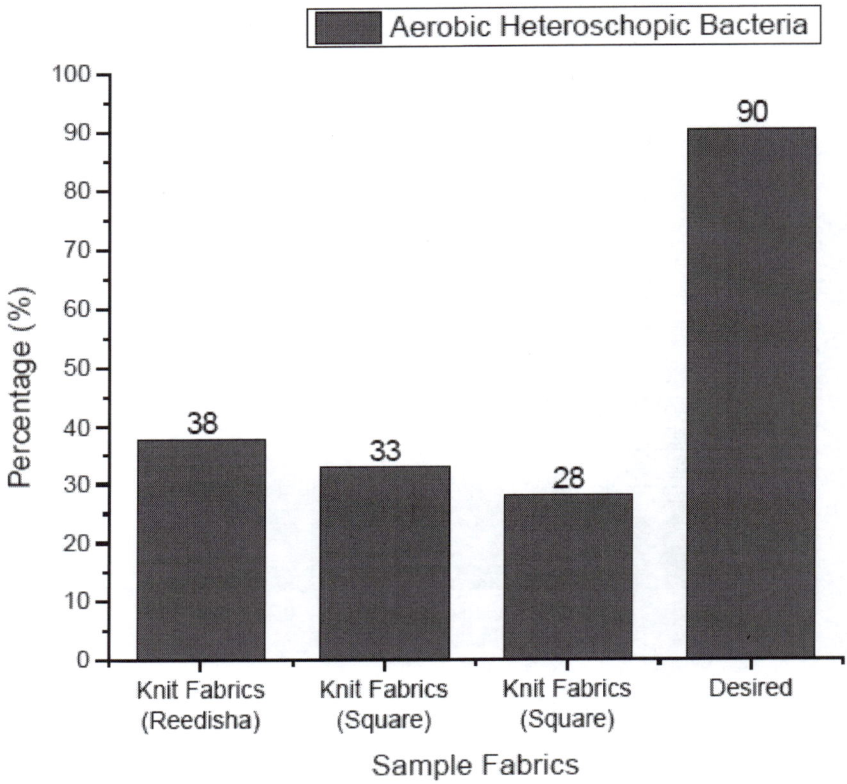

Figure 10. Assessment of antimicrobial agent used on export quality fabrics by Aerobic heterotrophic bacteria

3.4.2. Total Coliform

From Figure 11 it can be seen that, sample 1, 2, & 3 can prevent 22%, 18% and 22% total coliform respectively. The more the sample can inhibit the growth of bacteria means antimicrobial finish is successful. The desired rate of inhibition the growth of bacteria should be 90%.

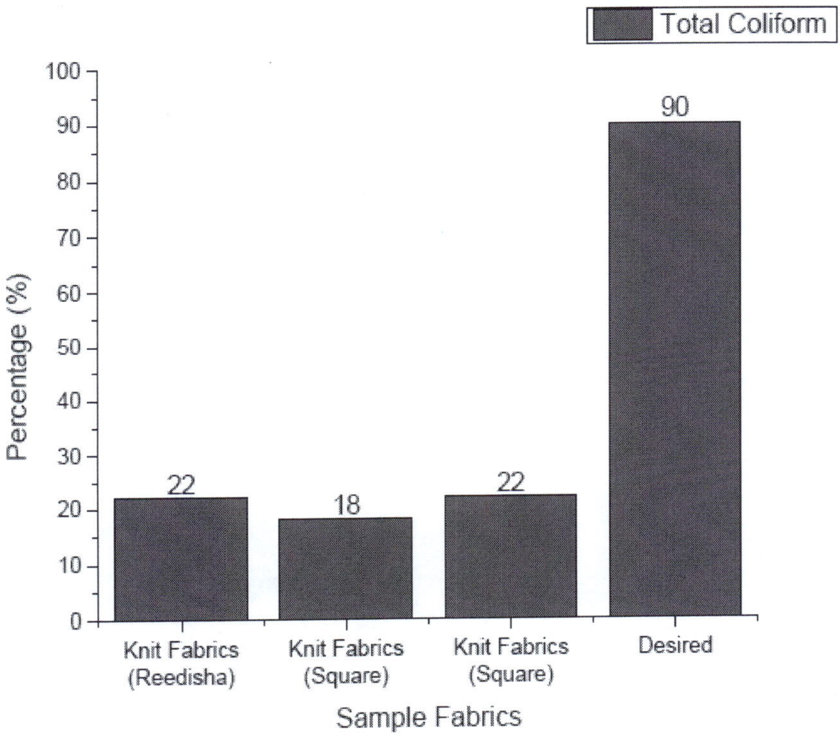

Figure 11. Assessment of antimicrobial agent used on export quality fabrics by Aerobic heterotrophic bacteria

3.4.3. Fecal Coliform

From Figure 12 it can be seen that, sample 1, 2, & 3 can prevent 16%, 14% and 18% fecal coliform respectively. The more the sample can inhibit the growth of bacteria means antimicrobial finish is successful. The desired rate of inhibition the growth of bacteria should be 90%.

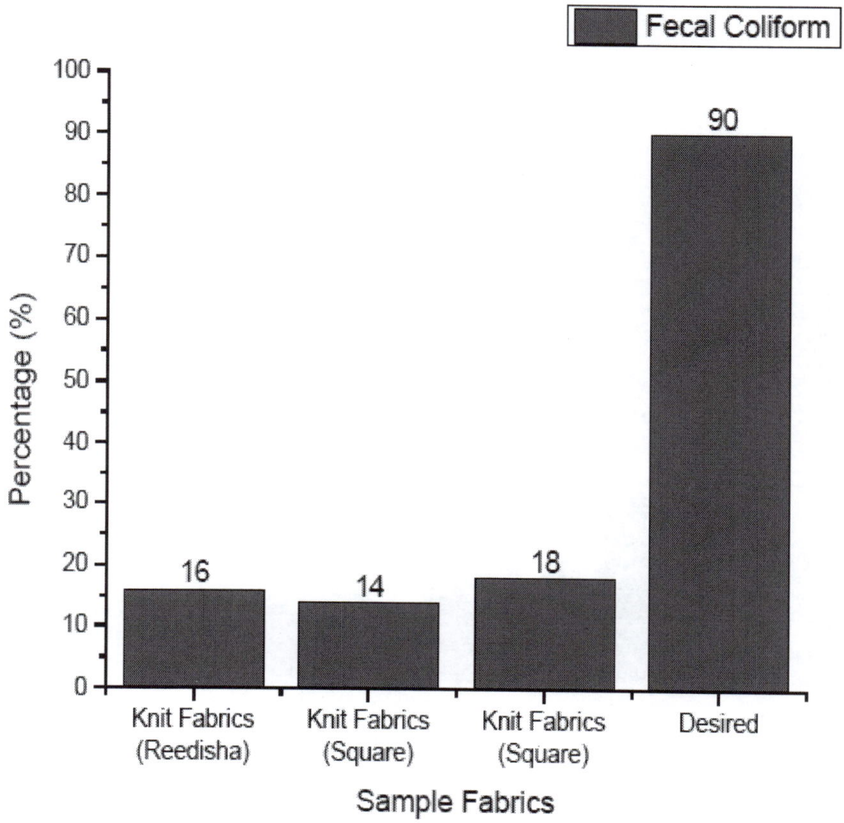

Figure 12. Assessment of antimicrobial agent used on export quality fabrics by Fecal Coliform bacteria

3.4.4. Shigella-Salmonellae

From Figure 13 it can be seen that, sample 1, 2, & 3 can prevent 14%, 12% and 18% Shigella-Salmonellae respectively. The more the sample can inhibit the growth of bacteria means antimicrobial finish is successful. The desired rate of inhibition the growth of bacteria should be 90%.

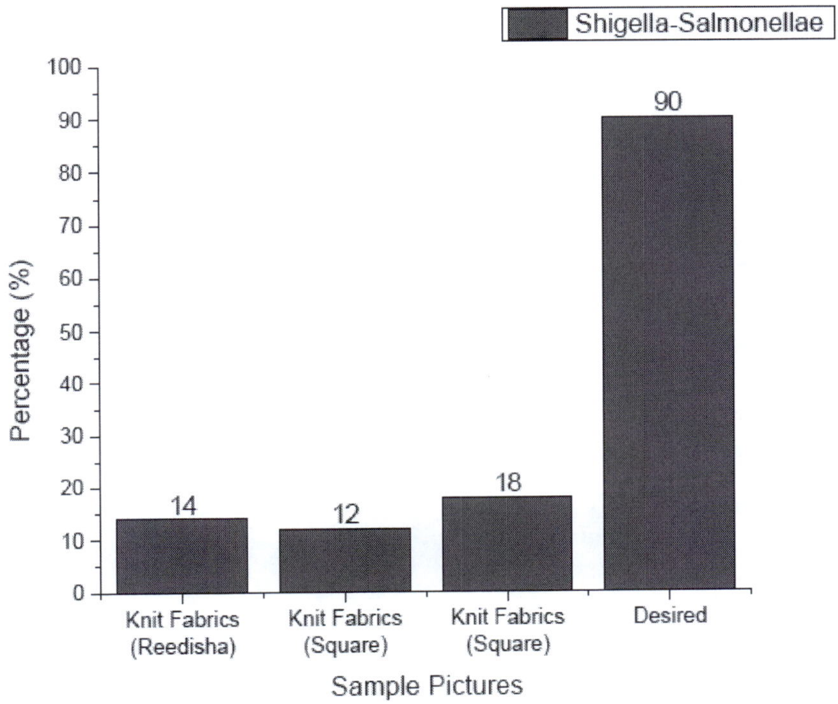

Figure 13. Assessment of antimicrobial agent used on export quality fabrics by Shigella-Salmonellae bacteria

3.4.5. Staphylococcus Aureus

From Figure 14 it can be seen that, sample 1, 2, & 3 can prevent 25%, 22% and 26% Staphylococcus Aureus respectively. The more the sample can inhibit the growth of bacteria means antimicrobial finish is successful. The desired rate of inhibition the growth of bacteria should be 90%.

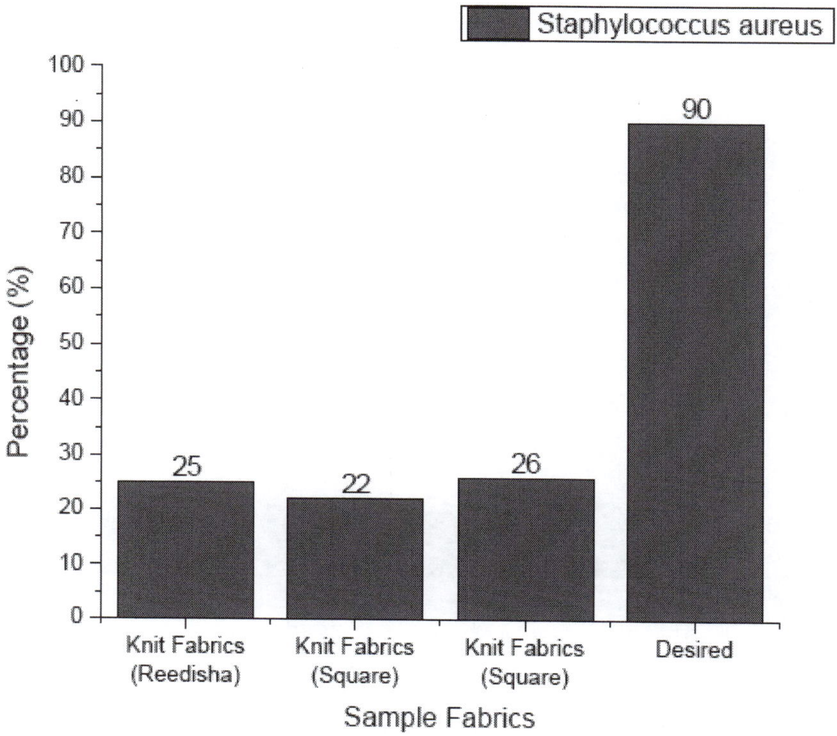

Figure 14. Assessment of antimicrobial agent used on export quality fabrics by Staphylococcus Aureus bacteria

3.4.6. Fungi Count

From Figure 15 it can be seen that, sample 1, 2, & 3 can prevent 30%, 35% and 28% Fungi respectively. The more the sample can inhibit the growth of bacteria means antimicrobial finish is successful. The desired rate of inhibition the growth of bacteria should be 90%.

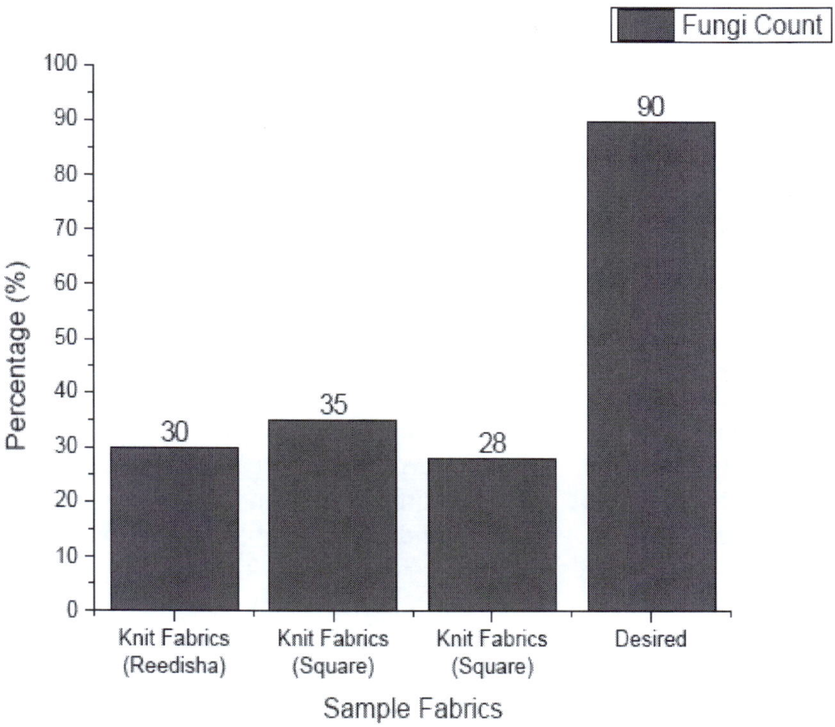

Figure 15. Assessment of antimicrobial agent used on export quality fabrics by Fungi Count

4. CONCLUSIONS

This paper mainly divided in to two major part of microbial quality assessment of the textiles used in medical sectors or hospitals and antimicrobial exportable finished fabrics. The tests performed to investigate those criteria include: Total Viable Bacteria Count (TVBC), Total Coliform Count (TCC), Total Fecal Coliform Count (TFCC), Total Staphylococcus Count (TSC), Total fungi Count (TFC).

In the result of microbiological assessment of medical textile, TFCC and Shigella-Salmonellae were found to be entirely absent.

Therefore, it may be concluded that the textiles used in medical sectors were good since they do not contain the most pathogenic bacteria Shigella-Salmonellae. A point is to be noted again that these types of bacteria are responsible for typhoid, dysentery, water and food borne diseases. The standard tolerance for Staphylococcus aureus should be 1.0 × 102, but in the test result it has been found that the minimum value of it is 1.0 × 104 and the maximum value is 9.0 × 103. The values found cross the standard limit. This type of bacteria is particularly responsible skin disease. The TCC was also found in the test results that cross the standard limit which is 3.0 cfu/gm only. The minimum value for TCC found is 1.8 × 103 and the maximum value of it found to be 2.4 × 103. The TCC bacteria are responsible for urinary tract infections, respiratory tract infections and other extra intestinal infections. Fungi were also found in the test result that cross the standard limits 0 (zero). The minimum value of it found to be 1.0 × 103 and maximum value 6.4 × 103. The fungi are large organisms and responsible for cancer and malnutrition. The research has found that the textiles used in medical or hospitals do not contain the most pathogenic bacteria Shigella-Salmonellae and total fecal coliform which are regarded to be very harmful. On the other hand the textile materials contain fungi types of microbes which are very harmful.

To identify the potency of antimicrobial finishing agent six types of tests have been carried out. The standard level of potency of antimicrobial agent should be 90% which although can be attained. In this work, the minimum efficacy for Aerobic heterotrophic bacteria has been found to be 28% and maximum 38%. The maximum efficacy for coliform has been found to 18% and maximum 28%. The maximum efficacy for fecal coliform has been found to be 14% and maximum 18%. The maximum and minimum values for staphylococcus have been found to be 22% and 26% respectively. The maximum and minimum values for Shigella-Salmonellae have been found to be 12% and 18% and for Fungi the values are 28% and 35% respectively. With reference to the results it may be concluded that collected antimicrobial finished fabric are able to inhibit the growth of microorganisms and their inhibition growth rate is not satisfactory.

ACKNOWLEDGEMENTS

This research work has been done in the microbiology laboratory of Primeasia University.

REFERENCES

1. H. Mucha, D. Hoter and M. Swerev 'Antimicrobial Finishes and Modifications' Melliand International, May 2002, vol. 8, p-148-151 and I. Home. 'Antimicrobials Impart Durable Finishes' International Dyer, December 2002, p-9-11.

2. New Multifunctional Textiles: Antimicrobial Treatments, 'Intelligent textile structures-application, production and testing' 12-13/5/2005, Thessaloniki, Greece, amphitheater of Thessaloniki Technology Park.

3. 'Antimicrobial Textiles: an Overview' Dr. T. Ramachandran, Member, K Rajendrakumar, Non-Member, and R. Rajendran, Non-Member.

4. H. Mucha, D. Hoter and M. Swerev, 'Antimicrobial Finishes and Modifications' Melliand International, May 2002, vol. 8, p-148-151, I Home 'Antimicrobials Impart Durable Finishes', International Dyer, December 2002, p-9-11 and D Gupta 'Antimicrobial Finishing of Textiles' www.resil.com.

5. Bauchop T. & Elsden S. R. 'The growth of microorganisms in relation to their energy supply' Journal of General Microbiology, 1960 and Tempest D. W. 'The biochemical significance of microbial growth yields: a reassessment TIBS' Elsevier/North Holland Biomedical Press, 1978, p-180.

6. Linko, M. 'An evaluation of enzymatic hydrolysis of cellulosic materials', in Advances in Biochemical Engineering, 5, 39, 1977, Bertran, M.S. and Dale, B.E. 'Enzymatic hydrolysis and recrystallization behavior of initially amorphous cellulose', Biotech. Bioeng, 27, 177, 1985 and Ghose, T.K. 'Cellulase biosynthesis and hydrolysis of cellulosic substances', in Advances in Biochemical Engineering, 6, 25, 1977.

7. 'Statistical Pocket Book of Bangladesh (1st Edition, 2007). Bangladesh Bureau of Statistics.

8. Mitchell A, Spencer M, Edmiston C, Role of healthcare apparel and other healthcare textiles in the transmission of pathogens: A review of the literature, Journal of Hospital Infection (2015),

9. Hawkins G, Stewart S, Blatchford O. Should HCWs be screened routinely for methicillinresistant Staphylococcus aureus? A review of the evidence. J Hosp Infect 2011;77:285–289.

10. Ibarra M, Flatt T, Van Maele D, Ahmed A, Fergie J, Purcell K. Prevalence of methicillinresistant Staphylococcus aureus nasal carriage in HCWs. Pediatr Infect Dis J 2008;27: 1109-1111.

11. Ferrero, F., et al., Sustainable antimicrobial finishing of cotton fabrics by chitosan UV-grafting: from laboratory experiments to semi industrial scale-up, Journal of Cleaner Production (2013),

7

CHAPTER

Photocatalytic Degradation of Azure A Using N-Doped Zinc Oxide

Priya Rathore*, Rakshit Ameta, Sanyogita Sharma

Department of Chemistry, PAHER University, Udaipur, India

ABSTRACT

Photocatalysis has been predicted as a promising technology for waste water treatment. N-doped zinc oxide has been used as an effective catalyst for carrying out number of chemical reactions, but limited work has been reported on use of N-doped ZnO as photocatalyst. In the present work, the photocatalytic degradation of Azure A was carried out in the presence of N-doped zinc oxide and the progress of the reaction was observed spectrophotometrically. The morphologies and structures of the as-synthesized nanomaterials were investigated by using FT-IR and DRS techniques. On the basis of observations, a tentative mechanism has been proposed for the photocatalytic degradation of dye.

Keywords: Azure A, Nanoparticles, N-Doped Zinc Oxide, Photocatalysis

1. INTRODUCTION

Water pollution is a serious threat for the environment. The sources of pollutants are generally textile and chemical industries using organic dyes. Therefore, methods for treatments of such polluted water are required. The waste water flows in the drains corrodes and incrusts the sewerage pipes. If allowed to flow in drains and rivers, it also effects the quality of drinking water in hand pumps making it unfit for human consumption. It leads to leakage in drains increasing their maintenance cost. Such polluted water provides breeding ground for bacteria and virus. All the organic materials present in the wastewater from a textile industry have a great concern in water treatment.

Dye degradation is a process, in which the large dye molecules are broken into smaller molecules like water, CO_2 and mineral byproducts. The waste water that the industry releases contains some amount of these dye molecules, which cause water pollution because many of them are non-reactive towards light, acids, bases and O_2. So the colour of the material becomes permanent.

Photocatalysis is the addition of light to semiconductor generally metal oxides and sulphides, which results in moving electrons from valence band to conduction band. The electron hole pairs formed will react with O_2 and H_2O molecules to form superoxides anions and $^{\cdot}OH$ radical, which have more oxidizing and reducing properties for the dye compounds used in many industries. Due to the importance of photocatalysis, researches are applying photocatalytic decomposition process to degrade the organic pollutants of industrial waste water before discharging it in mainstream.

Various metal oxide semiconductors, such as TiO_2, ZnO, WO_3 and SnO_2, have been studied and applied as photocatalysts. Among them, TiO_2 is considered the most efficient photocatalyst due to its abundant availability, cost-effectiveness and chemical stability. However, recent researches have shown that ZnO can be used as a more efficient photocatalyst compared with TiO_2 [1] [2] . An important characteristic is that ZnO contains various intrinsic defects. Introducing defects into ZnO structure can not only narrow the band gap to increase the solar light harvesting capability, but also bring many split energy gaps to set up a possible catalytic systems, which sense and shoot environmental contaminants [3] [4].

Kothari et al. [5] reported the photocatalytic degradation of evans blue, an azo dye and its mixture with amaranth in presence of ZnO. Photocatalytic mineralization of methylene blue using buoyant TiO_2-coated polystyrene beads was observed by Fabiyi and Skelton [6] . Ameta et al. [7] have used Iron(III) oxide for photobleaching of methylene blue, crystal violet and malachite green. Punjabi et al. [8] observed the reduction of Congo red by ascorbic acid and EDTA over CdS as a photocatalyst. Increase in photocatalytic activity of ZnO was reported by Benjamin et al. [9] by coating it with some natural pigments.

Egzar et al. [10] carried out the photocatalytic degradation of aniline blue dye using different semiconductors such as ZnO, ZnS and SnO_2. Photocatalytic degradation of various dyes i.e. orange-II, orange-G, Congo red, indigo carmine, crystal violet and malachite green has been reported by Hachem et al. [11] . Pare et al. [12] carried out the photocatalytic degradation of lissamine fast yellow in the aqueous suspension of ZnO under artificial light. The photocatalytic bleaching of eosin using ZnO and effect of surface charge was investigated by Vyas et al. [13] . Ameta et al. [14] carried out the photodegradation of naphthol green B dye using antimony trisulphide as a heterogeneous catalyst.

Attia [15] has reported the photocatalytic oxidation of iodide ion in the aqueous suspension of zinc oxide and its sensitization with riboflavin where as Habibi and Askari [16] used Mn-doped ZnOnanocomposite thin film coated on glass by a simple spin-coating method to degradeanazo textile dye. Ahmad et al. [17] synthesized Al-doped ZnO photocatalysts with different Al concentrations (0.5 - 6.0 mol%). Their results showed that this photocatalyst doped with 4.0 mol% Al exhibited five times enhanced photocatalytic activity compared to pure ZnO. Peng et al. [18] have prepared Ag-sensitized ZnO and compared the activity of ZnO and Ag/ZnO. They found that Ag/ZnO showed better photocatalytic performance under simulated solar light for the degradation of phenol and methyl orange.

2. MATERIAL AND METHOD

2.1. Preparation of ZnO Nanoparticles

Zinc acetate dihydrate [$Zn(CH_3COO)_2 \cdot 2H_2O$] (131.6 g) was added to 3 L of deionized waterunder vigorous stirring till homogeneous solution was obtained. Then 172.8 g of sodium dodecyl sulphate (SDS) and 0.2 M of sodium hydroxide were added into the above solution under continuous stirring till pH value reached 12 and heated at 160°C for 14 h in an oven. The precipitates were recovered by centrifugation and washed with deionized water and ethanol several times [19] . The product was dried at 60°C for 3 h and labelled as uncalcined zinc oxide (Znc). A portion of this product (10 g Znc) was calcined at 400°C for 4 h, cooled to room temperature, ground in agate mortarand labelled as calcined zinc oxide (Zc). At the end of the reaction, sodium dodecyl sulfate (SDS) and CH_3COONa were eliminated during washing of $Zn(OH)_2$. The final yield of ZnO was 90%.

2.2. Preparation of Nitrogen-Doped ZnO

The N-doped ZnO was synthesized in a solvent free reaction. 20 g of uncalcined zinc oxide was added to 40 g of urea and grinded in an agate mortar. The mixture was calcined in a ceramiccrucible at 400°C for 4 h and cooled [20] . The product was ground to fine powder and labeled as nitrogen-doped zinc oxide (NZ).

3. CHARACTERIZATION OF COMPOUND

3.1. Fourier Transform Infrared Spectroscopy (FT-IR)

FT-IR analysis of both the prepared samples were done using Bruker FT-IR. FT-IR results showed an absorption peak at 450 - 540 cm^{-1} and 1562 cm^{-1}, which confirmed the presence of ZnO nanoparticles [21] -[24] ; another intense, broad peak near 3360 cm^{-1} represented the hydrogen bonded O-H stretching vibration [25] . Strong peaks obtained at 1562 cm^{-1} and at 1404 cm^{-1}confirmed the presence of hydrolyzed particles.

3.2. Diffuse Reflectance Spectroscopy (DRS)

As-prepared pure ZnO and N-doped ZnO were analyzed by DRS to know the band gap. DRS analysis was done using UV Vis spectrophotometer with DRS, Model UV vis-3000+, Lab-India. The samples were run with the mixture of dry $BaSO_4$ at room temperature. It was found in DRS results that the band gap of pure ZnO was 3.4 eV and it was 3.2 eV for N-doped ZnO. It confirms the reduction of band gap by doping ZnO with nitrogen, as a consequence N-doped ZnO was found better in compare to pure ZnO for degradation of Azure A.

3.3. Scanning Electron Microscope (SEM)

SEM were recorded on SEM Jeol Japan make, 5610LV model of undoped as well as nitrogen doped Zinc Oxide (Figure 1 and Figure 2). It is clearly from micrograph that undoped Zinc Oxide was larger in size as compared to nitrogen doped Zinc Oxide. These, it may be concluded that nitrogen doping of Zinc Oxide reduces its particle size. It was substainated by XRD data where particle size of undoped ZnO is 45.31 nm and it was reduced to 29.16 nm after doping.

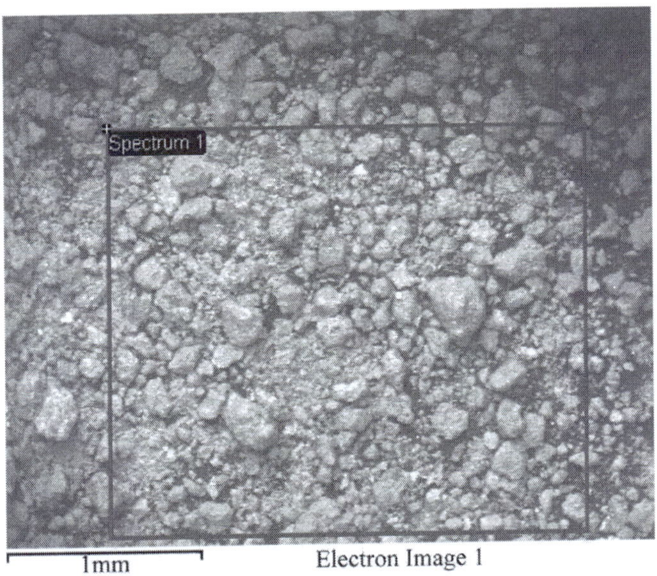

Figure 1. SEM undoped ZnO.

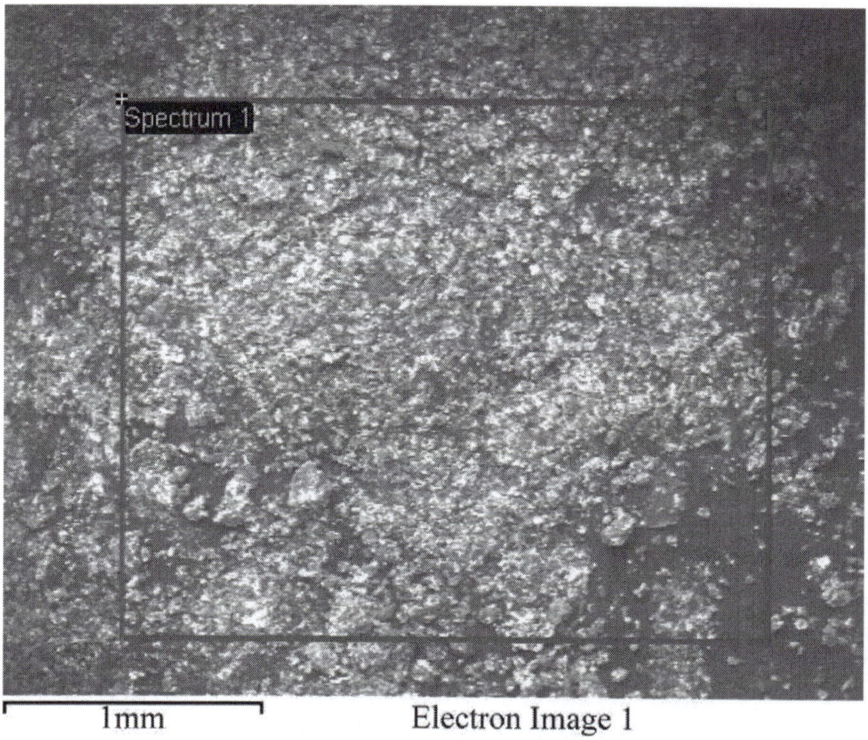

Figure 2. SEM nitrogen doped ZnO.

4. EXPERIMENT

Azure A dye and semiconducting N-doped zinc oxide powder were used in the present investigation of AR grade. All the solutions were prepared in doubly distilled water. The photocatalytic degradation of the dye was observed by taking dye solution and zinc oxide together. Irradiation was carried out by keeping the whole assembly exposed to a 200 W Tungsten lamp (Philips; light intensity = 50.0 $mW \cdot cm^{-2}$). The intensity of light at various distances from the lamp was measured with the help of a solarimeter. The pH of the solutions was measured with the help of digital pH-meter.

0.02918 g of Azure A was dissolved in 100 mL of doubly distilled water to prepare their 1.0×10^{-3} M solution, which was used as stock solution. The stock solution was further diluted as and when required. The absorbance of the dye solution were determined with the help of a spectrophotometer at λ_{max} = 620 nm for Azure A. The progress of reaction was observed by measuring absorbance of the reaction mixture at regular time interval during exposure. Decreasing trend of absorbance showed that dye was degraded during this process.

A decrease in absorbance of Azure A solution was observed with increasing time of exposure. A plot of 1 + log A against time was found linear for pure ZnO and nitrogen doped ZnO, which indicates that the degradation of Azure A follows pseudo-first order kinetics. The rate constant was measured with the help of Equation (1).

$$k = 2.303 \times \text{slope} \tag{1}$$

4.1. Azure A

Azure A is an organic compound with the chemical formula $C_{14}H_{14}ClN_3S$ (Figure 3). It is a light blue to dark blue dye. It is used for screening test for mucopolysaccharides. It is also known as Giemsa stain, Azure A chloride and dimethylthionine.

IUPAC name	(7-Amino-8-methyl-phenothiazin-3-ylidene)-dimethyl-ammoniumchloride
Molecular formula	$C_{14}H_{14}ClN_3S$
Molecular weight	291.80 g·mol^{-1}
Appearance	Green to dark brown powder
Solubility	Water

5. RESULTS AND DISCUSSION

Various factors, which affect degradation of Azure A dye have been studied to get the condition for optimum degradation rate.

5.1. Effect of pH

The effect of pH on the rate of degradation of N-doped ZnO was investigated in the pH range 6.0 to 10.0. The results are reported graphically in Figure 4.

[Azure A] = 1.6×10^{-5} M, N-doped ZnO = 0.10, Light intensity = 60.0 mW · cm^{-2}.

It has been observed that the rate of photocatalytic degradation of Azure A increases as pH was increased and it attained optimum value at pH 8.0. The rate was decreased when pH of the dye solution was further increased. It may be due to the fact that on increasing the pH of the solution, there is an attraction between cationic dye molecule and negatively charged ^{-}OH ions. But, at higher pH, the cationic dye molecules convert into neutral form and hence, there is less attraction between neutral dye molecule and negatively charged surface of semiconductor. As a result, the rate was retarded.

5.2. Effect of Azure A Concentration

The effect of dye concentration was observed by taking different concentrations of Azure A. The results are shown in Figure 5.

pH = 8.0, N-doped ZnO = 0.10 g, Light intensity = 60.0 mW · cm^{-2}.

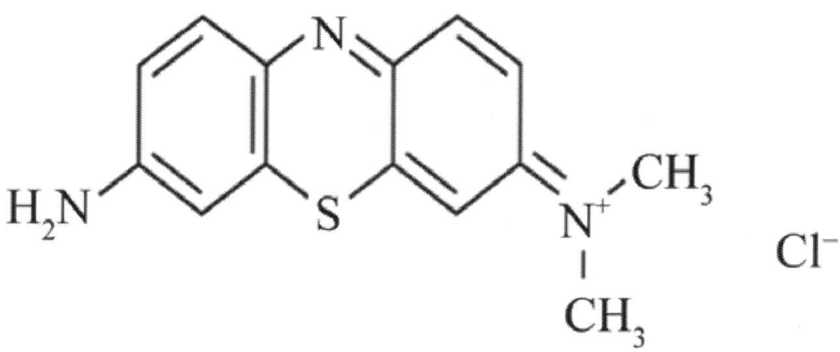

Figure 3. Structure of Azure A.

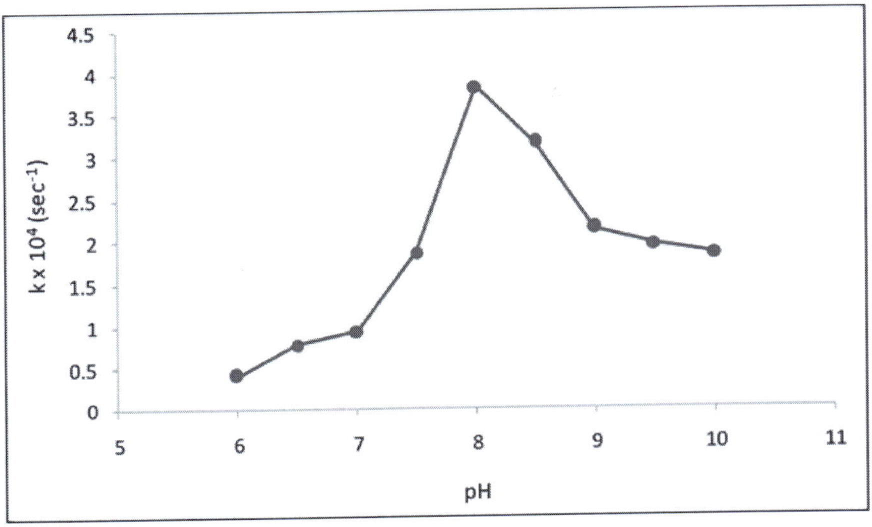

Figure 4. Effect of pH.

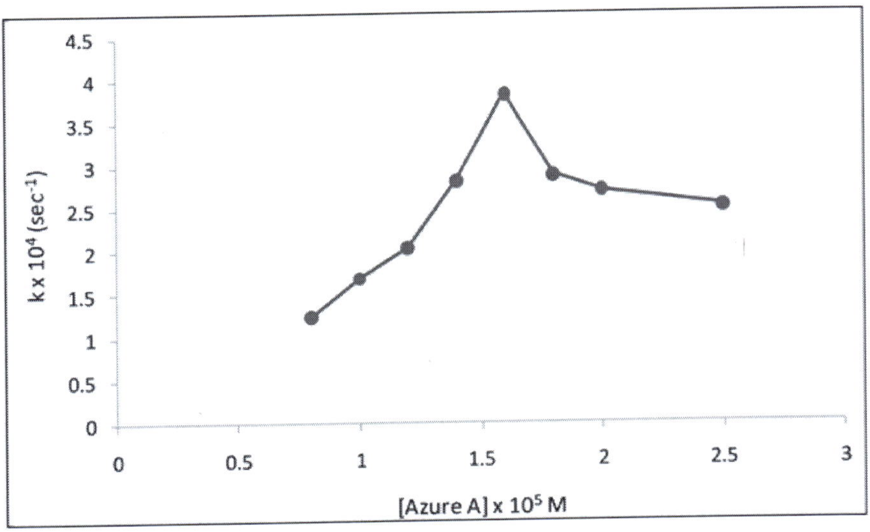

Figure 5. Effect of Azure A concentration.

It was observed that the rate of photocatalytic degradation of dye increases on increasing the concentration of Azure A upto 1.60×10^{-5} M. It may be attributed to the fact that as the concentration of the Azure A was increased, more dye molecules were available for excitation and consecutive energy/electron transfer and hence, an increase in the rate of degradation of the dye was observed. There was a decrease in degradation rate on increasing the concentration of dye above 1.6×10^{-5} M. This may be due to the fact that after a particular concentration, the dye may start acting as an internal filter and it will not permit the sufficient light intensity to reach the surface of the photocatalyst at the bottom of reaction vessel.

5.3. Effect of Amount of Semiconductor

The amount of semiconductor may also affect the degradation of dye and hence, different amounts of N-doped ZnO were used. The results are reported in Figure 6.

pH = 8.0, [Azure A] = 1.6×10^{-5} M, Light intensity = 60.0 mW · cm^{-2}.

The rate of reaction was found to increase on increasing the amount of semiconductor, N-doped zinc oxide. The rate of degradation reached to its optimum value at 0.10 g of the photocatalyst, because as the amount of the

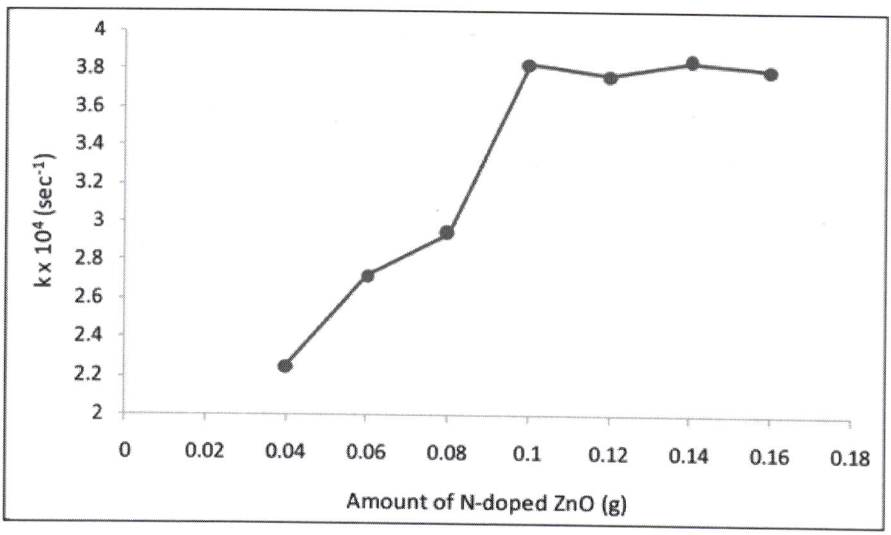

Figure 6. Effect of amount of N-doped ZnO.

semiconductor was increased, it provides more exposed surface area for degradation of dye molecules. Beyond 0.10 g, the rate of reaction becomes almost constant. After a particular value (0.10 g), an increase in the amount of semiconductor will only increase the thickness of layer of the semiconductor and not its exposed surface area. This was confirmed by taking reaction vessels of different sizes. It was observed that this point of saturation was shifted to a higher value for vessels of larger volumes while a reverse trend was observed for vessels of smaller capacities.

5.4. Effect of Light Intensity

The effect of light intensity on the photocatalytic degradation of Azure A was also investigated. The light intensity was varied by changing the distance between the light source and the exposed surface area of semiconductor. The results are given in Figure 7.

pH = 8.0, [Azure A] = 1.6×10^{-5} M, N-doped ZnO = 0.10 g.

These data indicate that photocatalytic degradation of Azure A was enhanced with the increase in intensity of light, because an increase in the light intensity will increase the number of photons striking per unit area per unit time of photocatalyst surface. There was a slight decrease in the rate of reaction as the intensity of light was increased beyond 60.0 $mW \cdot cm^{-2}$.

5.5. A Typical Run

After keeping the values of the above parameters constant for maximum rate constant, a typical run was observed for the photocatalytical degradation of the dye. The results are shown in Table 1 and graphically presented in Figure 8. Here, a comparative observation was made for pure ZnO and N-doped ZnO, which confirms that the rate was almost six times increased in the case of N-doped ZnO in compare to pure ZnO.

6. MECHANISM

On the basis of above observations, a tentative mechanism is proposed for the photocatalytic degradation of Azure A using N-doped ZnO semiconductor.

$^1Dye_0 \rightarrow {}^1Dye_1$ (Singlet excited state)

$^1Dye_1 \rightarrow {}^3Dye_1$ (Triplet excited state)

$SC \rightarrow e^- (CB) + h^+ (VB)$ or SC^*

$h^+ + {}^-OH \rightarrow {}^\bullet OH$

$HO^\bullet + {}^3Dye_1 \rightarrow$ Leuco dye

Leuco dye $\rightarrow$ Product

Azure A absorbs radiations of suitable wavelength and gives rise to its excited singlet state. Then it undergoes intersystem crossing (ISC) to give the triplet state of the dyes. On the other hand, the semi-conducting N-doped ZnO utilizes the radiant energy to excite its electron from valence band to the conduction band; thus, leaving behind a hole. This hole abstracts an electron from OH^- ions to generate $^\bullet OH$ radicals.

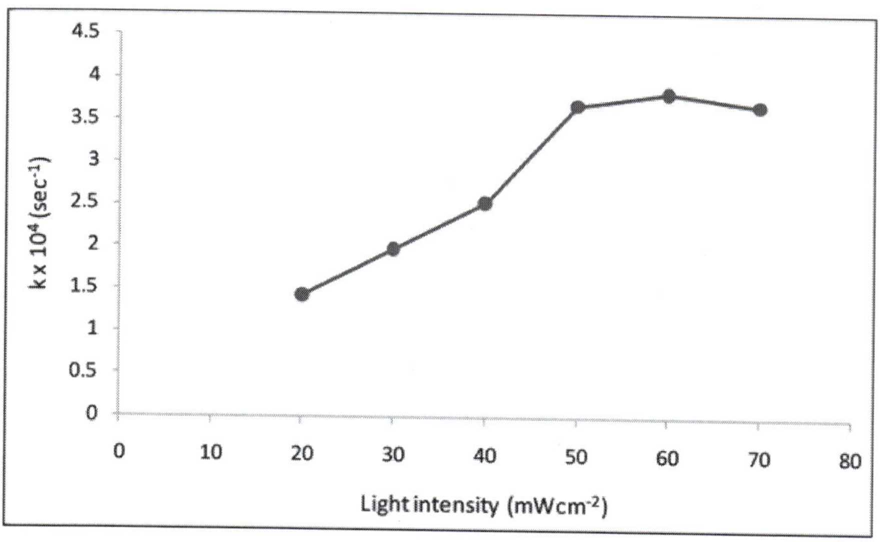

Figure 7. Effect of light intensity.

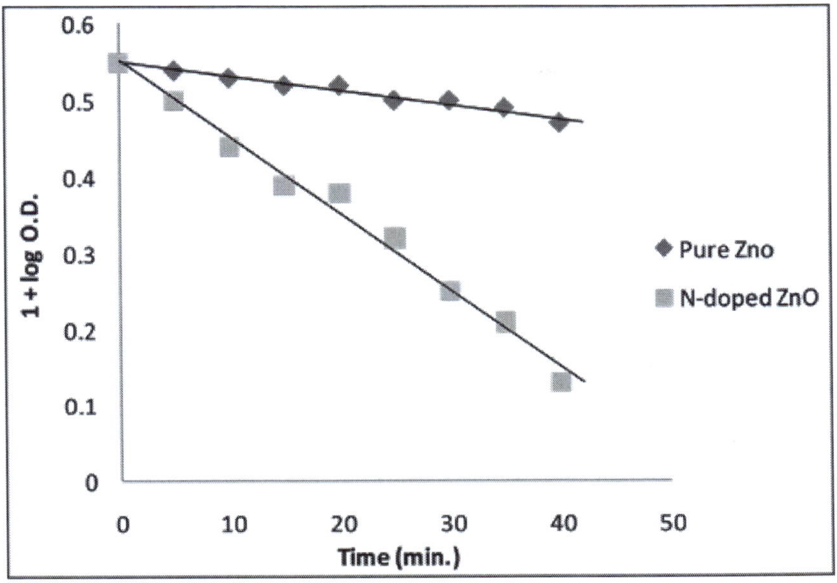

Figure 8. A typical run.

Table 1. A typical run.

Time (min.)	Pure ZnO (O.D.)	1 + log O.D.	N-doped ZnO (O.D.)	1 + log O.D.
0	0.356	0.55	0.356	0.55
5	0.347	0.54	0.318	0.50
10	0.339	0.53	0.278	0.44
15	0.337	0.52	0.249	0.39
20	0.333	0.52	0.241	0.38
25	0.322	0.50	0.213	0.32
30	0.317	0.50	0.178	0.25
35	0.300	0.49	0.163	0.21
40	0.301	0.47	0.138	0.13

pH = 8.0, [Azure A] = 1.6×10^{-5} M, N-Doped ZnO = 0.10 g, Light intensity = 60 $mW \cdot cm^{-2}$; Rate constant (k) for Pure ZnO = 0.60×10^{-4} sec^{-1} and for N-doped ZnO = 3.83×10^{-4} sec^{-1}.

These radicals further oxidize the Azure A dye to its leuco form, which may ultimately degrade to products. The participation of $^{\bullet}OH$ radicals as an active oxidizing species was confirmed by using hydroxyl radical scavenger 2-propanol, where the rate of degradation was drastically reduced.

7. CONCLUSION

Nitrogen doped zinc oxide was prepared by precipitation method by doping pure ZnO with urea and further the as-prepared samples were characterized by FT-IR, DRS and SEM techniques. Different rate affecting parameters like pH, dye concentrations, catalyst amount and light intensity were studied for the dye degradation. The observations revealed that Azure A dye could be degraded successfully by using N-doped ZnO under visible light. In comparative study, we found that the N-doped ZnO has shown six times enhanced photocatalytic activity than pure ZnO.

REFERENCES

1. Ahmad, M., Ahmad, E., Zhang, Y.W., Khalid, N.R., Xu, J.F., Ullah, M. and Hong, Z.L. (2013) Preparation of Highly Efficient Al-Doped ZnO Photocatalyst by Combustion Synthesis. Current Applied Physics, 13, 697-704.

2. Zhong, J.B., Li, J.Z., Xiao, Z.H., Hu, W., Zhou, X.B. and Zheng, X.W. (2013) Fabrication of Erbium-Doped Spherical-Like ZnO Hierarchical Nanostructures with Enhanced Visible Light-Driven Photocatalytic Activity. Materials Letters, 91, 1-4

3. Kamath, P.V., Huehn, R. and Nicolaesscu, R.A. (2002) A Sense and Shoot Approach for Photocatalytic Degradation of Organic Contaminants in Water. Journal of Physical Chemistry B, 106, 788-794.

4. Lin, B.X., Fu, Z.-X. and Jia, Y.-B. (2001) Green Luminescent Center in Undoped Zinc Oxide Films Deposited on Silicon Substrates. Applied Physics Letters, 79, 943-945.

5. Kothari, S., Ameta, P. and Ameta, R. (2007) Photocatalytic Bleaching of Evans Blue over ZnO Particulate System. Indian Journal of Chemistry, 46A, 432-435.

6. Fabiyi, M.E. and Skelton, R.L. (2000) Photocatalytic Mineralization of Methylene Blue Using Buoyant TiO2-Coated Polystyrene Beads. Journal of Photochemistry and Photobiology, 132A, 121-128.

7. Ameta, R., Vardia, J., Punjabi, P.B. and Ameta, S.C. (2006) Use of Semiconductor Iron(III) Oxides in Photocatalytic Bleaching of Some Dye. Indian Journal of Chemical Technology, 13, 114-118.

8. Punjabi, P.B., Ameta, R., Vyas, R. and Kothari, S. (2005) Photoreduction of Congo Red by Ascorbic Acid and EDTA over Cadmium Sulphide as Photocatalyst. Indian Journal of Chemistry, 44A, 2266-2269.

9. Benjamin, S., Vaya, D., Punjabi, P.B. and Ameta, S.C. (2011) Enhancing Photocatalytic Activity of Zinc Oxide by Coating with Some Natural Pigments. Arabian Journal of Chemistry, 4, 205-209.

10. Egzar, H.K., Mashkour, M.S. and Jude, A.M. (2013) Study the Photodegradation of Aniline Blue in Aqueous Phase by Using Different Photocatalysts. Asian Transactions on Basic & Applied Sciences, 3, 2221-4291.

11. Hachem, C., Bocquillon, F., Zahraa, O. and Bouchy, M. (2001) Decolourisation of Textile Industry Wastewater by the Photocatalytic Degradation Process. Dyes and Pigments, 49, 117-125. http://dx.doi.org/10.1016/S0143-7208(01)00014-6

12. Pare, B., Singh, P. and Jannalgadda, S.B. (2009) Artificial Light Assisted Photocatalytic Degradation of Lissamine Fast Yellow Dye in ZnO Suspension in Slurry Batch Reactor. Indian Journal of Chemistry, 48A, 1364-1369.

13. Vyas, R., Swarnkar, H. and Ameta, S.C. (2005) The Photocatalytic Bleaching of Eosin Using ZnO: Effect of Surface Charge. Chemical and Environmental Research, 14, 53-58.

14. Ameta, R., Punjabi, P.B. and Ameta, S.C. (2011) Photodegradation of Naphthol Green B in Presence of Semiconducting Antimony Trisulphide. Journal of the Serbian Chemical Society, 7, 1049-1055.

15. Attia, A.J. (2008) Photocatalytic Iodometry over Naked and Sensitized Zinc Oxide. National Journal of Chemistry, 32, 599-609.

16. Habibi, M.H. and Askari, E. (2011) Photocatalytic Degradation of an Azo Textile Dye with Manganese-Doped ZnO Nanoparticles Coated on Glass. Iranian Journal of Catalysis, 1, 41-44.

17. Ahmad, M., Ahmed, E., Zhang, Y., Khalid, N.R., Xu, J., Ullah, M. and Hong, Z. (2013) Preparation of Highly Efficient Al-Doped ZnO Photocatalyst by Combustion Synthesis. Current Applied Physics, 13, 697-704.

18. Peng, F., Zhu, H., Wang, H. and Yu, H. (2007) Preparation of Ag-Sensitized ZnO and Its Photocatalytic Performance under Simulated Solar Light. Korean Journal of Chemical Engineering, 24, 1022-1026.

19. Samaele, N., Amornpitoksuk, P. and Suwanboon, S. (2010) Effect of pH on the Morphology and Optical Properties of Modified ZnO Particles by SDS via a Precipitation Method. Powder Technology, 203, 243-247.

20. Zheng, M. and Wu, J. (2009) One-Step Synthesis of Nitrogen-Doped ZnO Nanocrystallites and Their Properties. Applied Surface Science, 255, 5656-5661

21. Maensiri, S., Laoki, P. and Promarak, V. (2006) Synthesis and Optical Properties of Nanocrystalline ZnO Powders by a Simple Method Using Zinc Acetate Dehydrate and Poly(Vinyl Pyrrolidine). Journal of Crystal Growth, 289, 102-106.

22. Bhatta, K.D., Sawant, D.N., Pinjari, D.V., Pandit, A.V. and Bhange, B.M. (2012) One Pot Green Synthesis of Nanosized Zinc Oxide by Sonochemical Method. Materials Letters, 77, 93-95.

23. Zamari, R., Zakaria, A., Ahangar, H.A., Darroudi, M., Zak, A. and Drummen, G.P.C. (2012) Aqueous Starch as a Stabilizer in Zinc Oxide Nanoparticle Synthesis via Laser Ablation. Journal of Alloys and Compounds, 516, 41-48.

24. Sankara, R.B., Venkatramana, R.S., Koteeswara, R.N. and Pramoda, K. (2013) Synthesis, Structural, Optical Properties and Antibacterial Activity of Co-Doped (Ag, Co) ZnO Nanoparticles. Research Journal of Material Sciences, 1, 11-23.

25. Seguel, G.V., Rivas, B.L. and Novas, C. (2005) Polymeric Ligand-Metal Acetate Interactions. Spectroscopic Study and Semi-Empirical Calculations. Journal of Alloys and Compounds, 50, 401-406.

8

CHAPTER

Study on the Changes of GSM (Gm/m2) of Grey Knitted Fabric from Pretreatment to Finishing

K. M. Faridul Hasan[1, 2, 3]

[1]*Department of Textile Engineering, Wuhan Textile University, Wuhan, Hubei, China*

[2]*Department of Textile Management & Business Studies, Bangladesh University of Textiles (BUTex), Dhaka, Bangladesh*

[3]*Department of Textile Engineering, Southeast University, Banani, Dhaka, Bangladesh*

ABSTRACT

This Project is on, Study on the changes of GSM (Gram per Square meter) of grey knitted fabrics from Pretreatment to finishing. This project helps to a greater extent to know about the whole procedures of knitted fabrics GSM Change from Pretreatment to finishing and compare with the theoretical and calculated values of GSM change & to take necessary initiatives against it. GSM change from Pretreatment to dyeing & finishing depends on the properties of fabric like structure of fabric, wales per inch, courses per inch, stitch length, yarn count, fabric spirality and shrinkage because the Fabric texture , extensibility of fabric depends on these property & also with the types of shade which is needed to be dyed like dark, medium or light shades. The final part of the project was to accumulate the data with sample and put them in table or chart as required as to identify & overcome the GSM related issues in a knit dyeing factory. It is seen that, average 15 to 30% GSM is changed from Grey stage to finishing stage in a knit dyeing factory.

Keywords: Fabric GSM, Pretreatment, Dyeing, Finishing

1. LITERATURE REVIEW

1.1. Fabric GSM

GSM means Gm per square meter .It's an essential property for knitted fabric. Due to fabric GSM the knitted fabric may be heavier or lighter. GSM varies from fabric to fabric, especially with the count & stitch length of fabric [1]. GSM is changed in Pretreatment, Dyeing & finishing to a greater extent. During these processes GSM Change depends on the types of chemicals & dyes used & also with the processes used [2, 3].

1.2. Some Terminology of Knitted fabrics

There are different types of related parameters are mentioned below [2-7]:

1. Wales per inch
2. Courses per inch
3. Stitch length
4. Yarn count
5. Yarn twist
6. Twist liveliness
7. Yarn type
8. Gram per square meter (GSM)
9. Spirality
10. Shrinkage
11. Area density
12. Fabric thickness

1. Wales per inch:

Wales per inch means how many numbers of wales are present in one inch, it is very important for calculation of knitted fabric GSM and fabric properties.

2. Courses per inch:

Courses per inch mean how many numbers of courses are present in one inch. It is also very important for calculation of knitted fabric GSM.

3. Stitch length:

Stitch length is theoretically is a single length of yarn which include one needle loop and half the length of Yarn (half of a sinker loop) between that needle loop and the adjacent needle loops on either side of it. Loop exists in coarse in course length and it is that which influence fabric dimension and other properties including weight.

4. Yarn count:

Yarn count is a numerical expression of fineness or coarseness of yarn.

Yarn count is calculated in two systems:

a. Direct system

b. Indirect system

5. Direct system:

In this system the count of yarn express the weight per unit's length of yarn. In this system, "higher the count, coarser the yarn" & "lower the count, finer the yarn". This system is used for thrown Polyester & other Synthetic yarn.

Count = (W X l)/ (w X L)

Here, W = the weight of sample

L= the length of sample

w = the unit weight in system

l= the unit length in system

System	Unit mass	Unit length	Uses
Tex	Gram	1000m	Polyester
Denier	Gram	9000m	Polyester

6. Indirect system:

In this system the count of yarn express the length per unit wt. In this system, "higher the yarn count, finer the yarn" & "Lower the count coarser the yarn. This system is generally used for cotton, worsted, woolen, linen etc and the yarn count calculation formula is.

1.3. GSM at a Glance

GSM depends on the fabric stitch length, course per unit length, wales per unit length and yarn count. GSM indicates the properties of the fabric such as fabric weight; in define area (one square meter). Fabric GSM calculations depend on the yarn count system. In direct system GSM calculation formula is given below [8-12]:

IN TEX SYSTEM:

GSM = WPI X CPI X Stitch length (mm) X 39.37 X 39.37 X TEX /(1000 X 1000)

Or, GSM = WPCm X CPCm x Stitch length (mm) X 100 X 100 X TEX / (1000X1000)

IN DENIER SYSTEM:

GSM = WPI X CPI X Stitch length X 39.37 X39.37 X Count in denier/(1000 x 9000)

Or GSM = WPCm X CPCm x Stitch length (mm) X 100 X 100 X Denier /(1000X9000)

In direct system GSM calculation formula is given below:

IN ENGLISH (Ne) SYSTEM:

GSM = WPI X CPI X Stitch length X 39.37 X39.37 X 453.6/(1000X840Xcount in NeX0.9144)

Table 1. Relation between yarn count, fabric type, stitch length and finished GSM

Fabric type	Yarn count	Stitch length	color	D X G	Finished dia	Finished GSM
Plain s/j	18s/1	2.94	white	26X24	30 inch	220-230
	20s/1	2.98	white	30X24	33.5 inch	200-210
	24 s/1	2.68	white	30X24	32 inch	175-185
	26s/1	2.66	white	30X24	31 inch	160-170
	28s/1	2.70	avg	26X24	25 inch	150-160
	30s/1	2.68	avg	30X24	30 inch	130-140
	34s/1	2.40	avg	26X24	24 inch	125-135
	40s/1	2.44	avg	24X24	20 inch	100-110
Single lacost	18s/1	3.00	avg	30X24	46 inch	245-255
	24s/1	2.64	avg	30X24	40 inch	210-215
	26s/1	2.60	avg	30X24	36inch	200-210
	30s/1	2.50	avg	30X24	33inch	180-190
1x1 Rib	26s/1	2.50	avg	40X18	41 inch	240
	24s/1	2.75	avg	30X18	64 inch	235
	26s/1	2.45	avg	40X18	40inch	245
	26s/1	2.55	avg	40X18	47inch	235
	26s/1	2.65	avg	32X18	32inch	230
	36s/1	2.65	avg	32X18	33inch	170
	24s/1	2.95	avg	34X18	32inch	220
	24s/1	2.90	avg	40X18	38inch	225
	28s/1	2.40	avg	30X24	38.5inch	232
2X1 L Rib	34s/1	2.60	avg	40X18	23inch	190
	34s/1	2.70	avg	32X18	20inch	185
2X2 Rib	40s/1	2.76	avg	30X20	35"T	160
	28s/1	2.87	avg	34X18	32"T	260
	26s/1	3.25	avg	34X18	31"T	205
	24s/1	3.50	avg	34X18	31.5"T	285
	20s/1	3.25	avg	34X20	55"(op)	260
2X2 L/Rib	20/1+40D	3.10	avg	30X18	18"T	390
	24/1+40D	2.90	avg	30X18	18"T	320
	30/1+70D	2.90	avg			320
	30/1+40D	2.98	avg			240
	30/1+40D	2.80	avg	30X18		240

Or GSM = WPCm X CPCm x Stitch length (mm) X 100 X 100 X 453.6/(1000Xcount in NeX0.9144)

IN METRIC SYSTEM:

GSM = WPI X CPI X Stitch length (mm) X 39.37 X 39.37 X1000/(1000X1000Xcount in Nm)

Or GSM = WPCm X CPCm x Stitch length (mm) X 100 X 100 X1000/(1000X1000X yarn count in Nm)

Other parameters are:

- Stitch length
- Fabric structure
- Finishing process
- Depth of shade
- Stitch density

Machine Gauge

NB: If the fabric is to be Enzyme washed, the stitch length should be kept (10%) less than the normal range. Because, enzyme wash reduces the total weight of the fabric by removing the floating fiber and hairy fiber [12-15].

For light color, the finished GSM varies near about 15-20% from grey GSM.

For average color, the finished GSM varies 20-25%.

For dark color, grey GSM is kept less (i.e. the stitch-length is large) because take up% of dark color is more and the weight of dyed fabric will be increased.

If the GSM varies 25-30%, it is not only necessary to control the VDQ pulley dia but also yarn tension & take up roller.

NB: We know that the stitch length increases with the increasing of VDQ pulley dia. For the change of VDQ pulley dia for one point stitch length will be changed approximately 0.04 mm (CAMBER, England, and Model: 95 QUATTRO II).

2. EXPERIMENTAL

2.1. Process Sequence

Normally a knit dyeing factory performs the dyeing of a fabric in following sequence. Here, pretreatment means scouring & bleaching of knit fabrics which is performed combinedly.

For Dyeing,

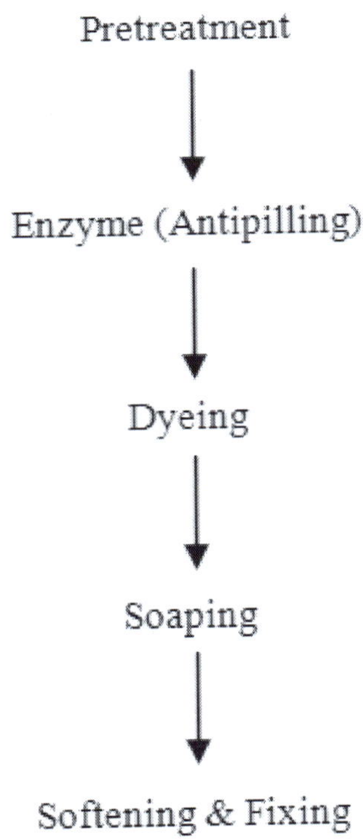

2.2. Chemicals & Dyes Used

Recipe-1 for 30-60°C to dye S/J fabric:

Process	Chemical /Dye	Concentration	Liquor Ratio
	Contipon S (10%)	2.50 g/l	7.20
	Jinsofter BSK	1.50 g/l	7.20
	Ablutex-AP-50(42%solⁿ)	0.75 g/l	7.20
PREBLEACHING CEL 110 & 95	Reduzin –AC-BL-100	1.50 g/l	7.20
	H_2O_2 (50%)	2.80 g/l	7.20
	Caustic Soda (pearls)	2.50 g/l	7.20
	Raduzin-THN (50% solⁿ)	2.40 g/l	6.50
	Acetic Acid	1.50 g/l	7.30
	SQ-117	0.75 g/l	6.80
	Retrocell-PLX-TRA	1.20 g/l	6.80
Antipilling	SQ-117	1.00 g/l	6.80
	Soda Ash	0.55 g/l	6.80
	Acetic Acid	0.25 g/l	6.80
	Jinlev- Cl-225	1.60 g/l	8.30
	Antisil Conz	0.30 g/l	8.30
	Jinsofter BSK	1.50 g/l	8.30
	Acetic Acid	.0 50 g/l	8.30
	Contipon S (10% solⁿ)	2.50 g/l	8.30
	Glauber salt	20.00g/l	8.30
Method 30-60°C	Levafix Brillian RED CA	0.27900%	6.30
	Levafix Blue-CA CA	0.00200%	6.30
	Levafix Amber- CA	0.01200%	6.30
	Soda Ash	.700g/l	8.30
	Soda Ash	6.30g/l	8.30
Soaping Extra Light	Acetic Acid	.50/l	6.8
Fixing & Softening	Acetic Acid	0.25g/l	8.30
	Bevasoft-CF (10% solⁿ)	2.00%	8.30

2.3. Recipe-2 for Bio Scouring Dyeing Elite to dye Lacoste Fabric

Process	Chemical /Dye	Concentration	Liquor Ratio
	Contipon S (10%)	2.50 g/l	6.9
	Jinsofter BSK	0.750 g/l	6.9
	Ablutex-AP-50(42%solⁿ)	0.75 g/l	6.9
PREBLEACHING CEL 110 &95	Reduzin –AC-BL-100	1.50 g/l	6.9
	H_2O_2 (50%)	2.80 g/l	6.9
	Caustic Soda (pearls)	2.50 g/l	6.9
	Raduzin-THN (50% solⁿ)	2.40 g/l	6.20
	Acetic Acid	1.50 g/l	7.00
	MA Scour AIN(Prep-IN)	1.50 g/l	8.00
	Glauber salt	20.00 g/l	8.00
	Remazol RED-3BS-A150	0.2090%	6.00
Bio scouring Dyeing Elite	Remazol Yellow-3BS	0.01703%%	6.00
	Remazol B.ELUE RSPL	0.47453%	6.00
	Soda Ash	4.00 g/l	8.00
	Green Acid(Jintex-CPS)	0.50 g/l	7.00
Soaping bioscouring	MA Disp-ARE(SOAP REA)	1.00 g/l	7.00
	Acetic Acid	0.25 g/l	8.00
Softening	Bevasoft-CF (10% solⁿ)	2.00 %	8.00

2.4. Recipe-3 for Iso Thermal Process

Process	Chemical /Dye	Concentration	Liquor Ratio
	Contipon S (10%)	2.50 g/l	7.20
	Jinsofter BSK	1.50 g/l	7.20
	Ablutex-AP-50(42%solⁿ)	0.75 g/l	7.20
PREBLEACHING CEL 110 &95	Reduzin –AC-BL-100	1.50 g/l	7.20
	H_2O_2 (50%)	2.80 g/l	7.20
	Caustic Soda (pearls)	2.50 g/l	7.20
	Raduzin-THN (50% solⁿ)	2.40 g/l	6.50
	Acetic Acid	1.50 g/l	7.30
	SQ-117	0.75 g/l	6.80
	Retrocell-PLX-TRA	1.20 g/l	6.80
Antipilling	SQ-117	1.00 g/l	6.80
	Soda Ash	0.55 g/l	6.80
	Acetic Acid	0.25 g/l	6.80
	Jinlev- Cl-225	1.60 g/l	8.30
	Antisil Conz	0.30 g/l	8.30
	Jinsofter BSK	1.50 g/l	8.30
	Acetic Acid	.0 50 g/l	8.30
	Contipon S (10% solⁿ)	2.50 g/l	8.30
	Glauber salt	20.00g/l	8.30
Method iso thermal	Novacron orange FNR	0.10230%	6.30
	Novacron B Red FN 3GL	0.01080%	6.30
	Cibacron FN Yellow FN- CA	0.18000%	6.30
	Soda Ash	.700g/l	8.30
	Soda Ash	6.30g/l	8.30
Soaping Extra Light	Acetic Acid	.50/l	6.8
Fixing & Softening	Acetic Acid	0.25g/l	8.30
	Bevasoft-CF (10% solⁿ)	2.00%	8.30

Machines & Equipment's Used:

- Dyeing M/C (Athena)
- Dewatering & Slitting m/c
- Stenter M/c
- Compacting M/c
- GSM Cutter
- Electric Balance

2.5. GSM Measurement

GSM is measured by using GSM cutter & Electric Balance. GSM means Gram per square meter. The size of the GSM cutter is equal to the area of 1 cm X1 Cm. The radius of GSM cutter is 6cm X 6cm.

2.6. Procedure of GSM Measurement

- To dry the sample properly.
- At first, the sample is cut by GSM cutter accurately.
- Then its weight is taken by the Electric Balance Before it, the balance is made zero.
- Then the obtained value is multiplied by 100 to get the required GSM.

E.g. If any sample weight is 1.80gm then its GSM would be 1.80 X100 = 180

2.7. Reasons of GSM Change

Pre Bleaching:

During this process Caustic Soda is used as scouring Agent to remove natural impurities, oil & wax from the fibre of fabric. Per oxide is used as bleaching agent to remove natural colour of the fibre. So the

GSM is reduced to a greater extent from Grey fabric as shown in table-2(b), 3(b), 4(b), 5(b), 6(b), 7(b), 8(b), 9(b), 10(b), 11(b), 12(b), 13(b), 14(b), 15(b).

Antipilling:

During this process enzyme is used as antipilling agent that removes the protruding fibre from the fabric. So the GSM is changed. It varies from fabric to fabric & the amount of protruding fibre present in the fabric as shown in table-2(b), 3(b), 4(b), 5(b), 6(b), 7(b), 8(b), 9(b), 10(b), 11(b), 12(b), 13(b), 14(b), 15(b).

Dyeing:

During dyeing significant amount of dyes are used depending on light, medium and dark shade. Here gluaber salt acts as electrolyte to penetrate the dye into fabric. Soda ash is used for fixation of dye into fabric by controlling the pH. So for fixing these dyes to a fabric considerable amount of GSM is increased depending on light, medium and dark shade. During this stage GSM is also increased for using lesser/larger amount of dye as shown in table-2(b), 3(b), 4(b), 5(b), 6(b), 7(b), 8(b), 9(b), 10(b), 11(b), 12(b), 13(b), 14(b), 15(b).

Soaping:

After dyeing there's some unfixed dye and also some other chemicals which are removed by soaping that's why the GSM is decreased as shown in table-2(b), 3(b), 4(b), 5(b), 6(b), 7(b), 8(b), 9(b), 10(b), 11(b), 12(b), 13(b), 14(b), 15(b).

Softening and fixing:

Here fixing agent and softening chemicals are used but GSM is not considerably changed as shown in table-2(b), 3(b), 4(b), 5(b), 6(b), 7(b), 8(b), 9(b), 10(b), 11(b), 12(b), 13(b), 14(b), 15(b).

Dewatering and slitting:

During dewatering for coloured fabric Inosoft CF & Jinsofter ME chemicals are used and for white fabric Jinsofter WCS & Microsol MA chemicals are used for why GSM is not considerably changed as shown in table-2(b), 3(b), 4(b), 5(b), 6(b), 7(b), 8(b), 9(b), 10(b), 11(b), 12(b), 13(b), 14(b), 15(b).

Stentering:

During stentering 10-15% GSM is increased by increasing overfeed % (up) and controlling overfeed % (down).GSM is changed as shown in table-2(b), 3(b), 4(b), 5(b), 6(b), 7(b), 8(b), 9(b), 10(b), 11(b), 12(b), 13(b), 14(b), 15(b) in this stage.

Compacting:

During compacting 10-15% GSM is increased by increasing overfeed % (up) as shown in table-2(b), 3(b), 4(b), 5(b), 6(b), 7(b), 8(b), 9(b), 10(b), 11(b), 12(b), 13(b), 14(b), 15(b).

By absorbing the humidity from the atmosphere 2-3% GSM is increased after compacting and buyer allow ±5%.

2.8. Description of Selected Fabric

Single Jersey:

This is produced by one set of needle only with all the loops intermeshed in the same direction. It's constructed with knit loop. Its face & back can be easily understood. It's used for T-shirt, fashionable knit garments, Underwear etc.

1×1 Rib:

Rib requires two sets of needle operating between each other. So wales of face & back stitches are knitted on each side of fabric. Rib fabric is used for waist bands, collar, Cuffs etc.

Double Lacost :

It's a Tuck-Knit single jersey structure. One set of needle is used for this structure.

GSM change For Single jersey fabric (First sample): During overall wet processing & finishing process of Single jersey fabric GSM is changed to a greater extent. GSM is mainly changed in Pretreatment, Dyeing, Stentering & in Compacting stages. During Pretreatment overall impurities are removed by using scouring & Bleaching Agent for why GSM is extremely reduced than grey fabric which is shown in table-2(b). During Antipilling Protruding fibres are removed by using Enzyme for why GSM is reduced. During Dyeing A lot of dyes are used for Combination shade for why GSM is increased. During soaping extra unfixed color is cut for why GSM is reduced. During Stentering GSM is reduced. During Compacting GSM is increased to a greater extent due to increase Overfeed%. After Compacting GSM is increased by absorbing moisture (3-5%).

Table 2(a). GSM change For Single jersey fabric

Fabric name	Buyer & Colour	M/c no.	Lot no.	Count	Brand of yarn	Treatment observation stage	Obtained GSM
Single Jersy	TEMA; Pink	Athena -1/1	03	40/1 Count	GTNIL-YRN	Grey stage	120
						Pre-treatment	108
						Enzyme	106
						Dyeing stage	119
						Soaping	118
						Fixing & Softening	120
						Stentering stage	109
						Compacting stage	138
						Final/finished stage	140

Table 2(b). GSM change from stage to stage

GSM Change in processing stages	Changing Value	Difference between Process to final stage GSM	Required value of GSM for finishing from process
Grey to Pretreatment	120-108=12 ↓	Grey to Finishing	140-120=20
Pretreatment to Antipilling/Enzyme	108-106=2 ↓	Pretreatment to Finishing	140-108=32
Antipilling to Dyeing	119 - 106=13 ↑	Antipilling to Finishing	140-106=34
Dyeing to Soaping	119-118=1 ↓	Dyeing to Finishing	140-119=21
Soaping to Fixing & Softening	120-118=2↑	Soaping to Finishing	140-118=22
		Softening & Fixing to Finishing	140-120=20
Fixing & Softening to Stentering	120-108 =1 2↓	Stentering to Finishing	140-109=31
Stentering to Compacting	138-109 = 29↑	Compacting to Finishing	140-138=2
Compacting to Finished stage	138 -140 =2 ↑	Finishing(Required) to Finishing (Actual)	140-140 =0

GSM change For Single jersey fabric (second sample): During overall wet processing & finishing process of Single jersey fabric GSM is changed to a greater extent. GSM is mainly changed in Pretreatment, Dyeing, Stentering & in Compacting stages. During Pretreatment overall impurities are removed by using scouring & Bleaching Agent for why GSM is extremely reduced than grey fabric which is shown in table-3(a) & 3(b). During Antipilling Protruding fibres are removed by using Enzyme for why GSM is reduced. During

Dyeing a lot of dyes are used for Combination shade for why GSM is increased. During soaping extra unfixed color is cut for why GSM is reduced. During Stentering GSM is reduced. During Compacting GSM is increased to a greater extent due to increase Overfeed%. After Compacting GSM is increased by absorbing moisture (3-5%).

Table 3(a). GSM change For Single jersey fabric

Fabric name	Buyer & Colour	M/c no.	Lot no.	Count	Brand of yarn	Treatment observation stage	Obtained GSM
Single Jersy	TEMA; Pink	Athena -1/1	03	40/1 Count	GTNIL-YRN	Grey stage	120
						Pre-treatment	106
						Enzyme	104
						Dyeing stage	120
						Soaping	118
						Fixing & Softening	120
						Stentering stage	113
						Compacting stage	138
						Final/finished stage	139

Table 3(b). GSM change from stage to stage

GSM Change in processing stages	Changing Value	Difference between Process to final stage GSM	Required value of GSM for finishing from process
Grey to Pretreatment	120-106=18 ↓	Grey to Finishing	140-120=20
Pretreatment to Antipilling/Enzyme	106-104=2 ↓	Pretreatment to Finishing	140-106=34
Antipilling to Dyeing	120 - 104=16 ↑	Antipilling to Finishing	140-104=36
Dyeing to Soaping	120-118=2 ↓	Dyeing to Finishing	140-120=20
Soaping to Fixing & Softening	121-120=1↑	Soaping to Finishing	140-118=22
		Softening & Fixing to Finishing	140-120=20
Finishing to Stentering	121-113 = 8↓	Stentering to Finishing	140-113=28
Stentering to Compacting	138-113 = 25↑	Compacting to Finishing	140-138=2
Compacting to Finished stage	139 -138 =1 ↑	Finishing(Required) to Finishing (Actual)	140-139 =1

GSM change For Single jersey fabric (third sample): This is another sample test result from pretreatment to finishing stage as above table -4 (a) & 4(b) of same single jersey fabric. During overall wet processing & finishing process of Single jersey fabric GSM is changed to a greater extent. GSM is mainly changed in Pretreatment, Dyeing, Stentering & in Compacting stages. During Pretreatment overall impurities are removed by using scouring & Bleaching Agent for why GSM is extremely reduced than grey fabric. During Antipilling Protruding fibres are removed by using Enzyme for why GSM is reduced. During Dyeing A lot of dyes are used for Combination shade for why GSM is increased. During soaping extra unfixed color is cut for why GSM is reduced. During Stentering GSM is reduced. During Compacting GSM is increased to a greater extent due to increase Overfeed%. GSM is increased after compacting by absorbing moisture (3-5%).

Table 4(a). GSM change For Single jersey fabric

Fabric name	Buyer & Colour	M/c no.	Lot no.	Count	Brand of yarn	Treatment observation stage	Obtained GSM
Single Jersy	TEMA; Pink	Athena -1/1	03	40/1 Count	GTNIL-YRN	Grey stage	120
						Pre-treatment	109
						Enzyme	107
						Dyeing stage	122
						Soaping	120
						Fixing & Softening	122
						Stentering stage	114
						Compacting stage	138
						Final/finished stage	139

Table 4(b). GSM change from stage to stage

GSM Change in processing stages	Changing Value	Difference between Process to final stage GSM	Required value of GSM for finishing from process
Grey to Pretreatment	120-109=11 ↓	Grey to Finishing	140-120=20
Pretreatment to Antipilling/Enzyme	109-107=2 ↓	Pretreatment to Finishing	140-107=33
Antipilling to Dyeing	120 - 107=13 ↑	Antipilling to Finishing	140-10=130
Dyeing to Soaping	122-120=2 ↓	Dyeing to Finishing	140-122=18
Soaping to Fixing & Softening	122-120=2↑	Soaping to Finishing	140-120=20
		Softening & Fixing to Finishing	140-122=18
Fixing & Softening to Stentering	122-114 = 8↓	Stentering to Finishing	140-114=26
Stentering to Compacting	138-114 = 24↑	Compacting to Finishing	140-138=2
Compacting to Finished stage	138 -139 =1 ↑	Finishing(Required) to Finishing (Actual)	140-139 =1

GSM change For Single jersey fabric (Fourth sample): During overall wet processing & finishing process of Single jersey fabric GSM is changed to a greater extent as shown in table-5(a) & 5 (b). GSM is mainly changed in Pretreatment, Dyeing, Stentering & in Compacting stages. During Pretreatment overall impurities are removed by using scouring & Bleaching Agent for why GSM is extremely reduced than grey fabric. During Antipilling Protruding fibres are removed by using Enzyme for why GSM is reduced. During Dyeing A lot of dyes are used for Combination shade for why GSM is increased. During soaping extra unfixed color is cut for why GSM is reduced. During Stentering GSM is reduced. During Compacting GSM is increased to a greater extent due to increase Overfeed%. After Compacting GSM is increased by absorbing moisture (3-5%).

Table 5(a). GSM change For Single jersey fabric

Fabric name	Buyer & Colour	M/c no.	Lot no.	Count	Brand of yarn	Treatment observation stage	Obtained GSM
Single Jersy	TEMA; Pink	Athena -1/1	03	40/1 Count	GTNIL-YRN	Grey stage	120
						Pre-treatment	108
						Enzyme	106
						Dyeing stage	119
						Soaping	118
						Fixing & Softening	120
						Stentering stage	108
						Compacting stage	140
						Final/finished stage	141

Table 5(b). GSM change from stage to stage

GSM Change in processing stages	Changing Value	Difference between Process to final stage GSM	Required value of GSM for finishing from process
Grey to Pretreatment	120-108=1 ↓	Grey to Finishing	140-120=20
Pretreatment to Antipilling/Enzyme	108-106=2 ↓	Pretreatment to Finishing	140-108=32
Antipilling to Dyeing	106 - 119=13 ↑	Antipilling to Finishing	140-106=34
Dyeing to Soaping	119-118=1 ↓	Dyeing to Finishing	140-119=21
Soaping to Fixing & Softening	120-118=2↑	Soaping to Finishing	140-118=22
		Softening & Fixing to Finishing	140-120=20
Finishing to Stentering	120-108 = 12↓	Stentering to Finishing	140-108=32
Stentering to Compacting	140-108= 22↑	Compacting to Finishing	140-140=0
Compacting to Finished stage	141 -140 =1 ↑	Finishing(Required) to Finishing (Actual)	140-141 =1

Comparison with all the samples for GSM change of Single jersey fabric: Comparing with all the four test sample it is found that the difference among all the separate samples GSM is very little, which could be negligible. So it is clear that my different test data results are near about similar as shown in bellow graph-1. So we can take these variation of data from different stages as experimentally approved data to assume any GSM related issues in different stages of knit dyeing factory.

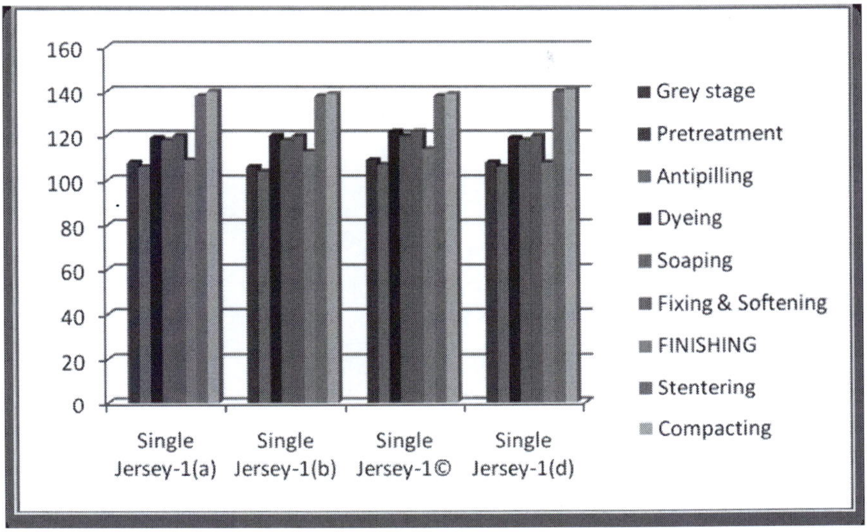

Figure 1. Graphical Representation of GSM Change

GSM change For Stripe fabric (First sample): During overall wet processing & finishing process of Stripe fabric GSM is changed to a greater extent as shown in table-6(a) & 6(b). GSM is mainly changed in Pretreatment, Dyeing, Stentering & in Compacting stages. During Pretreatment overall impurities are removed by using scouring & Bleaching Agent for why GSM is extremely reduced than grey fabric. During Antipilling Protruding fibres are removed by using Enzyme for why GSM is reduced. During Dyeing A lot of dyes are used for Combination shade for why GSM is increased. During soaping extra unfixed color is cut for why GSM is reduced. During Stentering GSM is reduced. During Compacting GSM is increased to a greater extent due to increase Overfeed%. After Compacting GSM is increased by absorbing moisture (3-5%).

Table 6(a). GSM change For Stripe fabric

Fabric name	Buyer & Colour	M/c no.	Lot no.	Count	Brand of yarn	Treatment observation stage	Obtained GSM
Stripe fabric	Grey Millange	HT-4/2	53	30/1	Paradise	Grey fabric	150
						Pre-treatment	145
						Enzyme	143
						Finishing	144
						Stentering	137
						Compacting	167
						Final/finished fabric	169

Table 6(b). GSM change from stage to stage

GSM Change in processing stages	Changing Value	Difference between Process to final stage of GSM	Required value of GSM for finishing from process
Grey to Pretreatment	150-145=05 ↓	Grey to Finishing	170-150=20
		Pretreatment to Finishing	170-145=25
Pretreatment to Antipilling/Enzyme	145-143=2 ↓	Enzyme to finishing	170-137=33
Enzyme to Finishing	144-143=1↑	Stentering to Finishing	170-137=33
Finishing to Stentering	144-137 = 7↓	Compacting to Finishing	170-167=3
Stentering to Compacting	167-137 = 30	to Finishing	170-169 =1
Compacting to Finished stage	169 -167 =2 ↑		

GSM change For Stripe fabric (Second sample): During overall wet processing & finishing process of Stripe fabric GSM is changed to a greater extent as shown in table-7(a) & 7 (b). GSM is mainly changed in Pretreatment, Dyeing, Stentering & in Compacting stages. During Pretreatment overall impurities are removed by using scouring & Bleaching Agent for why GSM is extremely reduced than grey fabric. During Antipilling Protruding fibres are removed by using Enzyme for why GSM is reduced. During Dyeing a lot of dyes are used for Combination shade for why GSM is increased. During soaping extra unfixed color is cut for why GSM is reduced. During Stentering GSM is reduced. During Compacting GSM is increased to agreater extent due to increase Overfeed%. After compacting GSM is increased by absorbing moisture (3-5%).

Table 7(a). GSM change For Stripe fabric

Fabric name	Buyer & Colour	M/c no.	Lot no.	Count	Brand of yarn	Treatment observation stage	Obtained GSM
Stripe fabric	Grey Millange	HT-4/2	53	30/1	Paradise	Grey fabric	150
						Pre-treatment	144
						Enzyme	143
						Finishing	144
						Stentering	137
						Compacting	168
						Final/finished fabric	169

Table 7(b). GSM change from stage to stage

GSM Change in processing stages	Changing Value	Difference between Process to final stage GSM	Required value of GSM for finishing from process
Grey to Pretreatment	150-144=06 ↓	Grey to Finishing	170-150=20
		Pretreatment to Finishing	170-145=25
Pretreatment to Antipilling/Enzyme	144-143=1 ↓	Enzyme to finishing	170-137=33
Enzyme to Finishing	144-143=1↑	Stentering to Finishing	170-137=33
Finishing to Stentering	144-137 = 7↓	Compacting to Finishing	170-167=3
Stentering to Compacting	168-137 = 29↓	to Finishing	170-169 =1
Compacting to Finished stage	169 -168 =1 ↑		

GSM change For Stripe fabric (Third sample): During overall wet processing & finishing process of Stripe fabric GSM is changed to a greater extent as shown in table-8(a) & 8(b). GSM is mainly changed in Pretreatment, Dyeing, Stentering & in Compacting stages. During Pretreatment overall impurities are removed by using scouring & Bleaching Agent for why GSM is extremely reduced than grey fabric. During Antipilling Protruding fibres are removed by using Enzyme for why GSM is reduced. During Dyeing A lot of dyes are used for Combination shade for why GSM is increased. During soaping extra unfixed color is cut for why GSM is reduced. During Stentering GSM is reduced. During Compacting GSM is increased to a greater extent due to increase Overfeed%. After Compacting GSM is increased by absorbing moisture (3-5%).

Table 8(a). GSM change For Stripe fabric

Fabric name	Buyer & Colour	M/c no.	Lot no.	Count	Brand of yarn	Treatment observation stage	Obtained GSM
Stripe fabric	Grey Millange	HT-4/2	53	30/1	Paradise	Grey fabric	150
						Pre-treatment	144
						Enzyme	143
						Finishing	144
						Stentering	138
						Compacting	168
						Final/finished fabric	169

Table 8(b). GSM change from stage to stage

GSM Change in processing stages	Changing Value	Difference between Process to final stage GSM	Required value of GSM for finishing from process
Grey to Pretreatment	150-144=06 ↓	Grey to Finishing	170-150=20
		Pretreatment to Finishing	170-145=25
Pretreatment to Antipilling/Enzyme	144-143=1 ↓	Enzyme to finishing	170-137=33
Enzyme to Finishing	144-143=1↑	Stentering to Finishing	170-137=33
Finishing to Stentering	144-138 = 6↓	Compacting to Finishing	170-167=3
Stentering to Compacting	168-138 =30↓	to Finishing	170-169 =1
Compacting to Finished stage	169 -168 =1 ↑		

Comparison with all the samples for GSM change of stripe fabric: Comparing with all the four test sample it is found that the difference among all the separate samples GSM is very little, which could be negligible. So it is clear that my different test data results are near about similar as shown in bellow graph-2. So we can take these variation of data from different stages as experimentally approved data to assume any GSM related issues in different stages of knit dyeing factory.

GSM change For Double Lacoste fabric (First sample): During overall wet processing & finishing process of Double Lacoste fabric, GSM is changed to a greater extent as shown in table-9(a) & 9(b). GSM is mainly changed in Pretreatment, Dyeing, Stentering & in Compacting stages. During Pretreatment overall impurities are removed by using scouring & Bleaching Agent for why GSM is extremely reduced than grey fabric. During Antipilling Protruding fibres are removed by using Enzyme for why GSM is reduced. During Dyeing A lot of dyes are used for Combination shade for why GSM is increased. During soaping extra unfixed color is cut for why GSM is reduced. During Stentering GSM is reduced. During Compacting GSM is increased to a greater extent due to increase Overfeed%. After Compacting GSM is increased by absorbing moisture (3-5%).

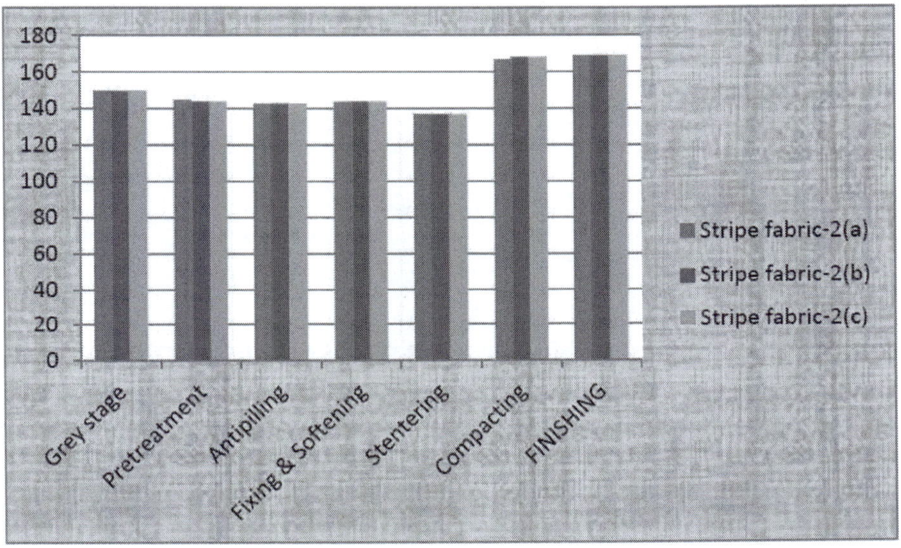

Figure 2. Graphical Representation of GSM Change

Table 9(a). GSM change For Double Lacoste Fabric

Fabric name	Buyer & Colour	M/c no.	Lot no.	Count	Brand of yarn	Treatment observation stage	Obtained GSM
Double Lacost	ZARA, Lavenda	HT-6	28	26/1	KAMYA-YRN	Grey Fabric	230
						Pre-treatment	221
						Dyeing	236
						Soaping	234
						Fixing & Softening	236
						Stentering	225
						Compacting	258
						Final/finished	260

Table 9(b). GSM change from stage to stage

GSM Change in processing stages	Changing Value	Difference between Process to final stage GSM	Required value of GSM for finishing from process
Grey to Pretreatment	230-221=9 ↓	Grey to Finishing	260-230=30
Pretreatment to Dyeing	236-221=15 ↓	Pretreatment to Finishing	260-221=39
Dyeing to Soaping	236-234=2 ↓	Dyeing to Finishing	260-236=24
Soaping to Fixing &Softening	236-234=2↑	Soaping to Finishing	260-234=26
		Fixing &Softening to Finishing	260-236=24
Finishing to Stentering	236-225= 11↓	Stentering to Finishing	140-108=32
Stentering to Compacting	258-225 = 33↑	Compacting to Finishing(Required)	140-140=0
Compacting to Finished stage	260-258 =2 ↑	Finishing(Required) to Finishing (Actual)	140-141 =1

GSM change For Double Lacoste fabric (Second sample): During overall wet processing & finishing process of Double Lacoste fabric, GSM is changed to a greater extent as shown in table-10(a) & 10(b). GSM is mainly changed in Pretreatment, Dyeing, Stentering & in Compacting stages. During Pretreatment overall impurities are removed by using scouring & Bleaching Agent for why GSM is extremely reduced than grey fabric. During Antipilling Protruding fibres are removed by using Enzyme for why GSM is reduced. During Dyeing a lot of dyes are used for Combination shade for why GSM is increased. During soaping extra unfixed color is cut for why GSM is reduced. During Stentering GSM is reduced. During Compacting GSM is increased to a greater extent due to increase Overfeed%. After Compacting GSM is increased by absorbing moisture (3-5%).

Table 10(a). GSM change For Double Lacoste Fabric

Fabric name	Buyer & Colour	M/c no.	Lot no.	Count	Brand of yarn	Treatment observation stage	Obtained GSM
						Grey Fabric	230
						Pre-treatment	223
Double Lacoste	ZARA; Lavenda	HT-6	28	26/1	KAMYA-YRN	Dyeing	236
						Soaping	234
						Fixing & Softening	236
						Stentering	223
						Compacting	259
						Final/finished	260

Table 10(b). GSM change from stage to stage

GSM Change in processing stages	Changing Value	Difference between Process to final stage GSM	Required value of GSM for finishing from process
Grey to Pretreatment	230-223=7 ↓	Grey to Finishing	260-230=30
Pretreatment to Dyeing	236-223=13 ↓	Pretreatment to Finishing	260-221=39
Dyeing to Soaping	236-234=2 ↓	Dyeing to Finishing	260-236=24
Soaping to Fixing &Softening	236-234=2↑	Soaping to Finishing	260-234=26
		Fixing &Softening to Finishing	260-236=24
Finishing to Stentering	236-225= 11↓	Stentering to Finishing	140-108=32
Stentering to Compacting	258-225 = 33↑	Compacting to Finishing (Required)	140-140=0
Compacting to Finished stage	260-258 =2 ↑	Finishing(Required) to Finishing (Actual)	140-141 =1

GSM change For Double Lacoste fabric (Third sample): During overall wet processing & finishing process of Double Lacoste fabric, GSM is changed to a greater extent as shown in table-11(a) & 11(b). GSM is mainly changed in Pretreatment, Dyeing, Stentering & in Compacting stages. During Pretreatment overall impurities are removed by using scouring & Bleaching Agent for why GSM is extremely reduced than grey fabric. During Antipilling Protruding fibres are removed by using Enzyme for why GSM is reduced. During Dyeing A lot of dyes are used for Combination shade for why GSM is increased. During soaping extra unfixed color is cut for why GSM is reduced. During Stentering GSM is reduced. During Compacting GSM is increased to a greater extent due to increase Overfeed% .After Compacting GSM is increased by absorbing moisture (3-5%).

Table 11(a). GSM change for Double Lacoste Fabric

Fabric name	Buyer & Colour	M/c no.	Lot no.	Count	Brand of yarn	Treatment observation stage	Obtained GSM
						Grey Fabric	230
						Pre-treatment	221
Double Lacoste	ZARA; Lavenda	HT-6	28	26/1	KAMYA-YRN	Dyeing	236
						Soaping	234
						Fixing & Softening	237
						Stentering	229
						Compacting	258
						Final/finished	260

Table 11(b). GSM change from stage to stage

GSM Change in processing stages	Changing Value	Difference between Process to final stage GSM	Required value of GSM for finishing from process
Grey to Pretreatment	230-221=9 ↓	Grey to Finishing	260-230=30
Pretreatment to Dyeing	236-221=15 ↓	Pretreatment to Finishing	260-221=39
Dyeing to Soaping	236-234=2 ↓	Dyeing to Finishing	260-236=24
Soaping to Fixing &Softening	237-234=3↑	Soaping to Finishing	260-234=26
		Fixing &Softening to Finishing	260-237=23
Fixing &Softening to Stentering	237-229= 8↓	Stentering to Finishing	260-229=31
Stentering to Compacting	258-225 = 33↑	Compacting to Finishing(Required)	260-258=2
Compacting to Finished stage	260-258 =2 ↑	Finishing(Required) to Finishing (Actual)	260-260 =2

GSM change For Double Lacoste fabric (Third sample): During overall wet processing & finishing process of Double Lacoste fabric, GSM is changed to a greater extent as shown in table-12(a) & 12(b). GSM is mainly changed in Pretreatment, Dyeing, Stentering & in Compacting stages. During Pretreatment overall impurities are removed by using scouring & Bleaching Agent for why GSM is extremely reduced than grey fabric. During Antipilling Protruding fibres are removed by using Enzyme for why GSM is reduced. During Dyeing A lot of dyes are used for Combination shade for why GSM is increased. During soaping extra unfixed color is cut for why GSM is reduced. During Stentering GSM is reduced. During Compacting GSM is increased to a greater extent due to increase Overfeed% .After Compacting GSM is increased by absorbing moisture (3-5%).

Table 12(a). GSM change For Double Lacoste Fabric

Fabric name	Buyer & Colour	M/c no.	Lot no.	Count	Brand of yarn	Treatment observation stage	Obtained GSM
						Grey Fabric	230
						Pre-treatment	221
Double Lacoste	ZARA; Lavenda	HT-6	28	26/1	KAMYA-YRN	Dyeing	236
						Soaping	234
						Fixing & Softening	235
						Stentering	225
						Compacting	258
						Final/finished	260

Table 12(b). GSM change from stage to stage

GSM Change in processing stages	Changing Value	Difference between Process to final stage GSM	
Grey to Pretreatment	230-221=9 ↓	Grey to Finishing	
Pretreatment to Dyeing	236-221=15 ↓	Pretreatment to Finishing	
Dyeing to Soaping	236-234=2 ↓	Dyeing to Finishing	
Soaping to Fixing &Softening	235-234=1↑	Soaping to Finishing	
		Fixing &Softening to Finishing	
Fixing &Softening to Stentering	235-225=10↓	Stentering to Finishing	
Stentering to Compacting	258-225 = 33↑	Compacting to Finishing(Required)	
Compacting to Finished stage	260-258 =2 ↑	Finishing(Required) to Finishing (Actual)	

Comparison with all the samples for GSM change of double Lacoste fabric: Comparing with all the four test sample it is found that the difference among all the separate samples GSM is very little, which could be negligible. So it is clear that my different test data results are near about similar as shown in bellow graph-1. So we can take these variation of data from different stages as experimentally approved data to assume any GSM related issues in different stages of knit dyeing factory.

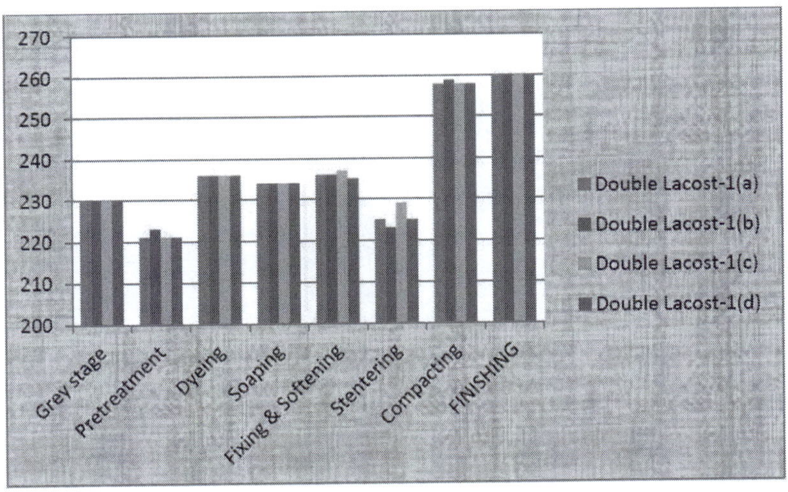

Figure 3. Graphical Representation of GSM Change:

GSM change For Double Lacoste fabric (1×1 Rib Fabric): During overall wet processing & finishing process of 1×1 Rib fabric, GSM is changed to a greater extent as shown in table-13(a) & 13(b). GSM is mainly changed in Pretreatment, Dyeing, Stentering & in Compacting stages. During Pretreatment overall impurities are removed by using scouring & Bleaching Agent for why GSM is extremely reduced than grey fabric. During Antipilling Protruding fibres are removed by using Enzyme for why GSM is reduced. During Dyeing A lot of dyes are used for Combination shade for why GSM is increased. During soaping extra unfixed color is cut for why GSM is reduced. During Stentering GSM is reduced. During Compacting GSM is increased to a greater extent due to increase Overfeed%. After Compacting GSM is increased by absorbing moisture (3-5%).

Table 13(a). GSM change for 1×1 Rib Fabric

Fabric name	Buyer & Colour	M/c no.	Lot no.	Count	Brand of yarn	Treatment observation stage	Obtained GSM
						Grey fabric	140
						Pre-treatment	133
						Enzyme	132
1×1 Rib	C &A;Mock Orange	Athena -4/1	36	28/1	Paradise	Dyeing	146
						Soaping	145
						Fixing & Softening	147
						Stentering	139
						Compacting	166
						Final/finished fabric	169

Table 13(b). GSM change from stage to stage

GSM Change in processing stages	Changing Value	Difference between Process to final stage GSM	Required value of GSM for finishing from process
Grey to Pretreatment	140-133=7 ↓	Grey to Finishing	170-140=30
Pretreatment to Antipilling/Enzyme	133-132=1 ↓	Pretreatment to Finishing	170-133=37
Antipilling to Dyeing	146 - 132=14 ↑	Antipilling to Finishing	170-132=38
Dyeing to Soaping	146 – 145=1 ↑↓	Dyeing to Finishing	170-146=24
Soaping to Fixing &Softening	145-147=2↑	Soaping to Finishing	170-145=25
		Fixing &Softening to Finishing	170-147=23
Fixing & Softening to Stentering	147-140 = 7↓	Stentering to Finishing	170-140=30
Stentering to Compacting	166-140 = 26↑	Compacting to Finishing(Required)	170-166=04
Compacting to Finished stage	166-169=3 ↑	Finishing(Required) to Finishing (Actual)	170-69 =1

GSM change For 1×1 Rib Fabric (Second sample): During overall wet processing & finishing process of 1×1 Rib fabric, GSM is changed to a greater extent as shown in table-14(a) & 14(b). GSM is mainly changed in Pretreatment, Dyeing, Stentering & in Compacting stages. During Pretreatment overall impurities are removed by using scouring & Bleaching Agent for why GSM is extremely reduced than grey fabric. During Antipilling Protruding fibres are removed by using Enzyme for why GSM is reduced. During Dyeing A lot of dyes are used for Combination shade for why GSM is increased. During soaping extra unfixed color is cut for why GSM is reduced. During Stentering GSM is reduced. During Compacting GSM is increased to a greater extent due to increase Overfeed%. After Compacting GSM is increased by absorbing moisture (3-5%).

Table 14(a). GSM change for **1×1 Rib** Fabric

Fabric name	Buyer & Colour	M/c no	Lot no.	Count	Brand of yarn	Treatment observation stage	Obtained GSM
						Grey fabric	140
						Pre-treatment	133
						Enzyme	132
1×1 Rib	C &A;Mock Orange	Athena -4/1	36	28/1	Paradise	Dyeing	147
						Soaping	145
						Fixing & Softening	147
						Stentering	138
						Compacting	166
						Final/finished fabric	169

Table 14(b). GSM change from stage to stage

GSM Change in processing stages	Changing Value	Difference between Process to final stage GSM	Required value of GSM for finishing from process
Grey to Pretreatment	140-133=7 ↓	Grey to Finishing	170-140=30
Pretreatment to Antipilling/Enzyme	133-132=1 ↓	Pretreatment to Finishing	170-133=37
Antipilling to Dyeing	147 - 132=15 ↑	Antipilling to Finishing	170-132=38
Dyeing to Soaping	146 – 145=1 ↑↓	Dyeing to Finishing	170-146=24
Soaping to Fixing &Softening	145-147=2↑	Soaping to Finishing	170-145=25
		Fixing &Softening to Finishing	170-147=23
Fixing & Softening to Stentering	147-138 = 9↓	Stentering to Finishing	170-140=30
Stentering to Compacting	166-138 = 28↑	Compacting to Finishing(Required)	170-166=04
Compacting to Finished stage	166-169=3 ↑	Finishing(Required) to Finishing (Actual)	170-69 =1

GSM change For 1×1 Rib Fabric (Second sample): During overall wet processing & finishing process of 1×1 Rib fabric, GSM is changed to a greater extent as shown in table-11(a) & 11(b). GSM is mainly changed in Pretreatment, Dyeing, Stentering & in Compacting stages. During Pretreatment overall impurities are removed by using scouring & Bleaching Agent for why GSM is extremely reduced than grey fabric. During Antipilling Protruding fibres are removed by using Enzyme for why GSM is reduced. During Dyeing A lot of dyes are used for Combination shade for why GSM is increased. During soaping extra unfixed color is cut for why GSM is reduced. During Stentering GSM is reduced. During Compacting GSM is increased to a greater extent due to increase Overfeed%. After Compacting GSM is increased by absorbing moisture (3-5%).

Comparison with all the samples for GSM change of 1×1 Rib Fabric: Comparing with all the four test sample it is found that the difference among all the separate samples GSM is very little, which could be negligible. So it is clear that my different test data results are near about similar as shown in bellow graph-1. So we can take these variation of data from different stages as experimentally approved data to assume any GSM related issues in different stages of knit dyeing factory.

Table 15(a). GSM change for 1×1 Rib Fabric

Fabric name	Buyer & Colour	M/c no.	Lot no.	Count	Brand of yarn	Treatment observation stage	Obtained GSM
						Grey fabric	140
						Pre-treatment	131
						Enzyme	133
1×1 Rib	C &A;Mock Orange	Athena -4/1	36	28/1	Paradise	Dyeing	146
						Soaping	145
						Fixing & Softening	147
						Stentering	139
						Compacting	168
						Final/finished fabric	169

Table 15(b). GSM change from stage to stage

GSM Change in processing stages	Changing Value	Difference between Process to final stage GSM	Required value of GSM for finishing from process
Grey to Pretreatment	140-131=9 ↓	Grey to Finishing	170-140=30
Pretreatment to Antipilling/Enzyme	131-133=2 ↓	Pretreatment to Finishing	170-131=39
Antipilling to Dyeing	146 - 133=13 ↑	Antipilling to Finishing	170-133=37
Dyeing to Soaping	146 – 145=1 ↑↓	Dyeing to Finishing	170-146=24
Soaping to Fixing &Softening	145-147=2↑	Soaping to Finishing	170-145=25
		Fixing &Softening to Finishing	170-147=23
Fixing & Softening to Stentering	147-140 = 7↓	Stentering to Finishing	170-140=30
Stentering to Compacting	147-139 = 8↑	Compacting to Finishing(Required)	170-168=02
Compacting to Finished stage	168-139=29 ↑	Finishing(Required) to Finishing (Actual)	170-169 =1

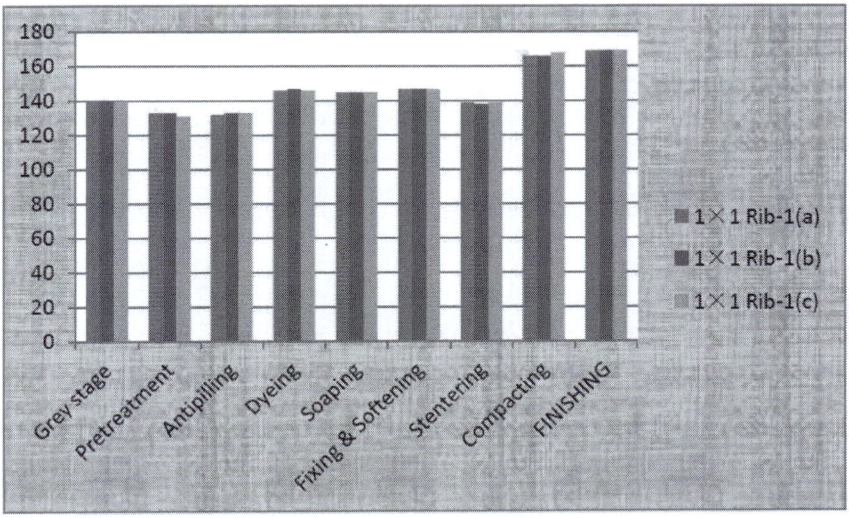

Figure 4. Graphical Representation of GSM Change

3. EFFECT OF GSM ON FINISHED FABRIC

√ 10-15% GSM of grey knit fabric is increased after overall Dyeing & finishing process.

√ If GSM is increased than diameter is decreased.

√ With the change of fabric GSM Shrinkage & Spirality of fabric is changed.

√ Fabric remains heavier or lighter with the change of GSM.

√ If GSM is increased in uncontrolled way than the fabric may be damaged due to extra overfeed which is responsible for GSM.

3.1. Cares that should be Taken for GSM Measurement

√ To collect two or more sample from different parts of the fabric of same lot & same order no. to get accurate result.

√ To ensure the sample is accurately dried. To ensure that no pressure is applied to the sample before cutting by the GSM cutter.

√ To measure the GSM before absorbing the moisture to get accurate result.

√ To ensure that the sample is cut by GSM cutter accurately.

- To ensure that the blade is sharp enough to cut the sample for GSM.
- To make sure that the Rubber pad is plane but not rough.
- To apply uniform Tension.
- To ensure that there's not any yarn of sample to be cut later.
- Numbering of sample for specific process is to be done.

√ To measure the GSM accurately.

- To clean the Balance properly
- To make zero the Electric balance reading.
- When the sample is put in the balance its covered with the head.
- The reading is taken & multiplied by 100

3.2. Faults that may Arise during GSM Measurement

- GSM may vary due to existing dirt variation in different parts of the fabric Grey Stage).
- If Pretreatment is not proper in all parts of the fabric.
- If GSM of Uneven shade (fabric) is taken.
- To take improper Weight.
- To cut the sample improperly.
- To take the Stentering sample which is not properly gripped?
- To take the measurement after long time

- If the fiber is immature in some particular places.
- If the Electric balance exist any dirt or dust.
- If the sample is remained rough through any pressure applied directly or indirectly to it before cutting by GSM cutter.

4. CONCLUSIONS

I tried my best to find out the Change of GSM of different fabrics (mentioned in the experimental data) from Pretreatment to finishing both in dyeing & finishing section.

For this purpose, I had to collect the sample of each & every processes for GSM measurement. For some fabrics we took more than one or two sample for accuracy of the result.

Through this measurement of GSM, I got the GSM of different fabrics for different processes & I was able to compare these GSM values for different fabric for different processes.

All most all the GSM values found are near about to acceptable limit. When we observed more than one GSM values for same fabric & same processes we did not get larger difference.

More research and development is required to monitor the Change of GSM in different processes.

REFERENCES

1. E.P.G. Ghol & L.D. Vilensky, Textile Science (1983), CBS Publishers & Distributors, 41-42.
2. S. Gordon and Y-L. Hsies, Cotton: Science and technology (2007), Woodhead Publishing Ltd, 3-5.
3. A.D. Broadbent, Basic Principle of Textile Coloration9201), SDC, 72.

4. E.R. Troatman, Dyeing & Chemical technology of Textile Fibers (1975), Charles Giffin & Company Ltd, 44.

5. Postle, R. (1990) "Fabric Objective Measurement Technology: Present Status and Future Potential", International Journal of Clothing Science and Technology, Vol. 2 No: 3, pp.7 – 17

6. Ukponmwan J. O. (1988). Correlation Between Objective and Subjective Methods of Assessing Fabric Handle, Journal of Textile Institute, No 4, 1988, pp 580-587.

7. Kawabata S. (1989). Fabric Performance in Clothing and Clothing Manufacture, Journal of Textile Institute, Vol 80, No 1, pp 19-50

8. Chen, Y., and Collier, B. J, (1997). Characterising Fabric End Use by Fabric Physical Properties, Textile Research Journal, Vol 67, No 4, pp 247-252.

9. BS (1979). Methods of Test for Textiles (British Standard Handbook No. 11), British Standards Institution, Revised Edition.

10. Gong R.H. and Mukhopadhyay S.K. (1993). Fabric Objective Measurement: A Comparative Study of Fabric Characteristics, Journal of Textile Institute, Vol 84, No 2, pp 192-198.

11. Roshan L. S. (1995), Importance of Mechanical and Physical Properties of Fabrics in the Clothing Manufacturing Process. International Journal of Clothing Science and Technology, Vol 7, No 2/ 3, pp 35-42.

12. Prof. V.A. shenai, Technology of Textile Processing (Vol II): Chemistry of Dyes and Principles of Dyeing (1993), Sevak Publications, 519-522.

13. John shore, Cellulosic Dyeing (19950, SDC, 212-213.

14. John Shore, Colorants and Auxiliaries (Vol II): Chemistry of Dyes and Principles of Dyeing (1993), Sevak publications, 519-522.

15. Prof. V.A. shenai, Technology of Textile Processing (Vol II): Chemistry of Dyes and Principles of Dyeing (1993), Sevak Publications, 519-523.

16. Hassan B. A. (2003). Effect of Mechanical and Physical Properties on Fabric Hand. Wood Head Publishing Limited, New York, pp 447 - 447.

9

CHAPTER

A Revolution of Silk Dyeing with FL Based Cotton-Reactive Dyes

Ratul Rehman [1, 2], Sakil Mahmud [1, 3], Md Ahsan Habib [1, 3], Ashraful Islam [1, 3]

[1]M. Sc. in Textile Chemistry and Dyeing & Finishing Engineering, School of Chemistry & Chemical Engineering, Wuhan Textile University, Hubei, China

[2]B. Sc. in Knitwear Manufacturing Technology, BGMEA Institute of Fashion & Technology, Dhaka, Bangladesh

[3]B.Sc. in Textile Engineering (Wet Processing Technology), School of Science & Engineering, Southeast University, Dhaka, Bangladesh

ABSTRACT

Silk is the royal fibre and its coloration is also an art. There are wide ranges of dyestuff accessible for silk dyeing. Nearly every type of dyestuff used for cotton and/or wool can be used for dyeing silk. In overall the dyestuffs are useful by techniques similar to those of wool or cotton dyeing. Dyeing of silk with acid dyes required more than boiling temperature, resulting fiber damage and more energy. In this study, silk fabric was dyed with three commercial dyes. There are Lanasol reactive dyes (which are manufactured specially for protein fibers like silk & wool dyeing) then Liyuansol reactive dye (fluorine based dye which is specially manufactured for cotton dyeing) and finally, Lanacron acid dyes (which is commercially available for protein dyeing). The first and foremost aim of this study is to introduce fluorine based cotton-reactive dyes in silk under boiling temperate. On the basis of a number of experiments, influences of temperature, sodium sulfate consumption, time duration, sodium carbonate dosage and different pH were disclosed.

Through orthogonal experiment, the optimum conditions of silk dyeing with above defined three dyes are defined. Results revealed that, fluorine based cotton reactive Liyuansol dyes has excellent or equal application performance at 90°C dyeing temperature like the dyes that is specially manufactured for protein fiber (Lanasol) or specially manufactured acid dye (Lanacron).

Keywords: Reactive dyes, Acid dyes, Silk dyeing, Fastness, Protein fiber dyeing, Fluorine based reactive dyes

1. INTRODUCTION

Silk is one of the most popular protein fiber extruded from animal. Because of its unique properties like wearing comfort, soft handle, good air permeability & elegant appearance its makes as the symbol of royalty [1]. It has been used as textile materials for a decade due to its excellences in super lustered, comfort, brilliant color shades obtained in dyeing & printing, tensile strength and so on. Although it containing all technical properties of fiber like strength, elasticity, resilience, drape ability, heat conductivity, absorbency, cleanliness & wash ability, shrinkage, effect of heat & light, resistance to mildew & insects and dye affinity but due to its serious inferiority in function performance, crease recovery, colorfastness to wash, dry and/or wet rub fastness, pholoyellowing resistance, colorfastness to perspiration, washing & wearing properties still limits its conventional uses in textile [2]. According to the customer demand, it is the right time for process the challenging goods of silk with outstanding features [3]. Up to present research, some researchers show in this field to improve the properties of fiber & fabric.

This paper explains the major components of silk fiber, acid dyes and reactive dyes. This paper also enclosed with some evidence on the silk properties, silk chemistry and about acid & reactive dyes chemistry, methods of silk with Fluorine (Fl) based reactive dye which is specially designed for cotton dyeing & an acid dyes which is usually used for protein fiber. Finally, comparing the results between silk dyeing with acid & reactive dyes as well as fruitful discussions was presented. The fast & foremost aim of this research work is due to define the unique & even color dyeing process of silk dyeing with fluorine based cotton-reactive dyes. Of course dyeing performance is strongly prejudiced by the time, appropriate pH, dye bath concentration, fixing time, alkali dosage, amount of salt dosage as well as promoter consumption. The efficient dyeing principle was computed by altering the important variable parameters on dyeing equilibrium.

2. LITERATURE REVIEW

Protein is a sensitive fiber. Even dyeing with acid dyes at the boil caused some degradation, particularly on the prolonged boiling to promote dye migration. This causes hydrolysis of various proteins [4]. The dyeing of silk with acid dye is uneven even though the distribution of the dye liquor is perfect. Found the problems like (1) undyed, (2) particularly or ring dyed, (3) completely but light dyed, (4) completely but heavily dyed [5]. To overcome the above-mentioned problems dyeing should be carried out at boil for period of 1-8 hours, under a variety of pH (3-7). Dyeing at boil more than 1 hour is essentially unnecessary. But the bulk shade is 10% weaker than the laboratory matching or previously obtained bulk samples. So, it is required to topping additional dyes. It is expected that the future research in this field will be the automatic color matching, dyeing weighting and dispensing, and automatic dye bath control [6].

3. SILK REVIEW

3.1. Chemical Composition of Silk

Various types of chemical components are composed in the chemical structure of the silk fiber. Here, I have presented a chemical composition of raw silk fiber. Fibroin (75%), Ash of Silk Fibroin (0.5%), Sericin (22.5%), Fat and Wax (1.5%), Mineral Salt (0.5%). This composition is calculated in 100%. It is important to know that, the fibroin is composed of 16 various amino acids. Raw Silk is composed of two filaments which are produced by the fibroin. These two filaments are adjusted by the gum which is known as sericin and different types of mineral salt, fat and wax also involved with it. This raw silk is treated for remove the gummy materials and various types of impurities from the silk by scouring process. Then the silk fiber is collected and supplied for producing silk fabric by weaving process [7].

3.2. Chemical Constituents of Silk

Silk is a natural protein, like wool fiber, due to this, mechanism of dyeing silk is dependent not only on free amino and carboxyl groups but also on phenolic with accessible –OH group. Because of slightly cationic character of silk with isoelectric point at above pH 5.0, it can be dyed with anionic dye such as acid, metal complexes, reactive and selected direct dyes. But the main objective of coloration of a textile fiber is that the permanency of the color and should not allow damage of natural abstract of fiber. This implies that it should not destroy its color during processing following coloration and dyeing & subsequent useful life (i. e. washing, light, rubbing, perspiration, and saliva) [8]. So whatever dyestuff we use for silk dyeing it is very essential to have permanency of that dyestuff. In this study there are three major components, which are silk fiber, acid dyes and reactive dyes. This paper starts with some information on the silk fiber and about acid dyes and re active dyes for silk are discussed. Finally, the results and discussions are presented [9].

3.3. Chemical Properties of Silk

Chemical properties of the silk fiber are given below [10]:

1. Effect of acids and bases: Acids and alkalis cause hydrolysis of polypeptide chain present in the fibroin fiber. Acid hydrolysis causes more damage to fiber than alkaline hydrolysis. Acid hydrolysis attacks at nearly all the peptide linkages as compared to alkali hydrolysis which occurs at the end of the peptide chain. Concentrated sulphuric acid and hydrochloric acid will dissolve the fiber and nitric acid result in change of color of silk. Dilute acid do not attack the fiber under mild condition. Hot caustic alkalis readily dissolve the

fiber. Weak alkalis attack fibroin when the treatment time at boiling point is prolonged.

2. Effect of organic solvents: Organic solvents are used as dry cleaning solvents and they do not dissolve the silk fiber.

3. Effect of oxidizing agents: Oxidation reactions are considered to take place at the side chains of tyrosine, the amino terminal residues of the main chains and at the peptide bonds. Silk fiber is susceptible to oxidizing agents. Therefore care is required during bleaching. Hydrogen peroxide and some peracids are used in bleaching of raw silk.

4. Effect of reducing agents: Fibroin resists the reducing agents such as hydrosulphite, sulphurous acid and its salt.

5. Color-fastness: The luster of silk will cause its dyed and printed silk textile materials to appear much brighter in color.

3.4. Physical Properties of Silk

There have a lot of physical properties of silk. Followings are the physical properties of silk:

- Color: The color of silk fiber could be yellow, brown, green or grey.
- Tensile Strength: Cotton is a strong fiber. It has a tenacity of 3.5 – 5 gm/den. The strength is greatly affected by moisture; the wet strength of silk is 75 – 85%, which is higher than dry strength.
- Elongation at break: 20 -25% at break.
- Elastic Recovery: Not so good.
- Specific Gravity: Specific gravity is 1.25 to 1.34.
- Moisture Regain (MR %): Standard moisture regain is 11% but can absorb up to 35%.

- Effect of Heat: Silk will withstand at higher temperatures than wool. It will remain unaffected for prolonged periods at 140C. Silk decompose at 175C.
- Effect of Sun Light: Sun light tends to encourage the decomposition of silk by atmospheric oxygen.
- Luster: Bright.

4. DYES REVIEW

4.1. Acid Dyes

Acid dyes are usually sodium salts of sulphonic acids, or less frequently of carboxylic acids, and are therefore anionic in aqueous solution. They will dye fibers with cationic sites. These are usually substituted ammonium groups in fibers such as wool silk & nylon. These fibers absorb acids. The acids can protonates the fibers amino groups, so they become cationic. Dyeing involves exchange of the anion associated with an ammonium ion in the fiber with a dye anion in the bath [4]

$\text{Fiber-NH}_2\ (s) + H^+\ (aq) + HSO_4^-\ (aq) \rightarrow \text{Fiber-NH}_3^+\ HSO_4^-\ (s)$

$\text{Fiber-NH}_3^+\ HSO_4^-\ (s) + \text{Dye-SO}_3^- \rightarrow \text{Fiber-NH}_3^+\ \text{Dye-SO}_3\ (s) + HSO_4^-\ (aq)$

An acid dye should have highly solubility in water and gives anion in acidic aqueous solution and which can react with cationic parts of the fibers. An acid dyes should have higher molecular weight in the range 300-1000 g/mole. The dyes with larger molecules have higher substantively for wool and nylon. Such dyes have slower diffusion in the fiber and therefore less ability to migrate and dye level. The more hydrophobic, high molecular weight dyes therefore have better fastness to the wet process. Anionic (Acid) dyes include many compounds from the most varied classes of dyes, which exhibit characteristic differences in structure (e.g., azoic, anthraquinone, triphenylmethane, and nitro dyes) but possess as a common feature water-solubilizing, ionic substituents. Sulphonated azo (acid) dyes constitute the major group and are mainly mono-azo & bis-azo compounds ranging in color from yellow, through red to violate & brown. There are some navy blue bis-azo dyes that can build up to gives black.

4.2. Reactive Dyes

Reactive dyes have proven to be one of the most successful classes of modern synthetic dyes. The reasons for this success lie in their shade versatility, their flexibility in application and the all-round good fastness properties obtained from dyeings produced with reactive dyes on wool, silk, cotton and regenerated cellulosic fibers [11].

Reactive dyes may be loosely defined as chromophores which contain pendant groups capable of forming covalent bonds with nucleophilic sites in fibrous substrates. Providing these covalent bonds are stable to conditions encountered in laundering, the wash-fastness properties of the colorations will be outstanding [11].

Reactive dye contain reactive group that allows fiber to react with dye by nucleophilic addition or substitution. One single dye can contain more than one reactive group (Figure 1). Difluorochloropyrimidine reacts with cellulose by nucleophilic substitution of labile chlorine, fluorine leaving group activated by adjacent nitrogen in a heterocyclic ring. In the other hand vinylsulphone group reacts with the cellulose by nucleophilic addition to a carbon-carbon double bond, usually activated by an adjacent electron attracting sulphone group (Figure 1). The sodium sulfonate groups make the dyes water soluable. Also $-NH_2$; -OH; $-SO_3^-$; -N=N- can form covalent bonds with the cellulose containing –OH group [4].

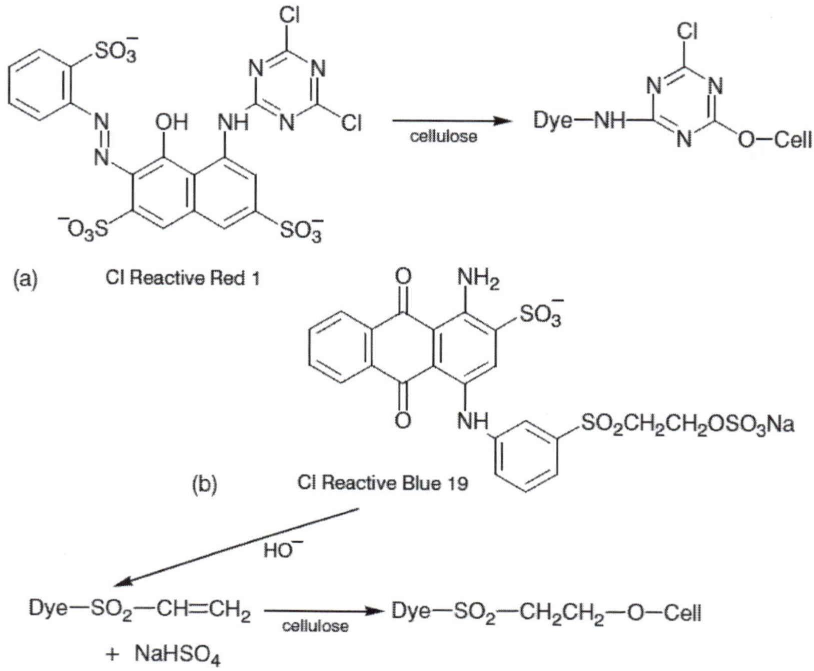

Figure 1. Nucleophilic substitution (a) and addition reactions (b) of reactive dues with cellulose

5. DYEING OF SILK

5.1. Dyeing of Silk with Acid Dyes

Dissolution of dyes in aqueous solvent produces a colored anion,

$$\underset{\text{Sodium salt of acid dye}}{A\text{-}SO_3Na} \xrightarrow{H_2O} \underset{\text{Colored Anion}}{A\text{-}SO_3^- + Na^+}$$

The protein and polyamide fibers produce cationic sites in water under acidic conditions, as the acidity of the solution is increased more cationic sites are produced under these strongly acidic conditions. These cationic sites are thus available for the acid dye anions to combine with through hydrogen bonding, vander waals forces or ionic bonding. These linkages are strong enough to break, and thus dyeing produced is fast. These contain acid groups, such as –COOH and –SO_3H which form attractions to the slightly basic –NH groups in the amide links of wool, silk and nylon [12].

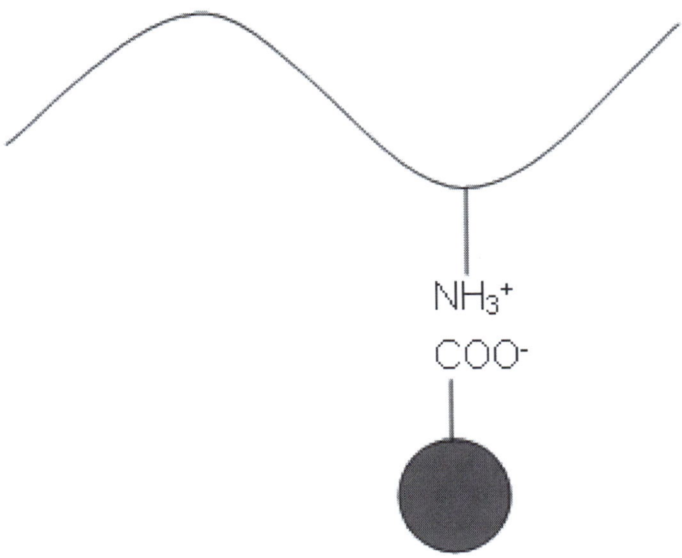

Figure 1. How Acid dye attached with silk

5.2. Dyeing of Silk with Reactive Dye

Exhaustion and fixation by a one-step process in an acidic or a neutral bath was the mode of dyeing used in silk & wool dyeing. Like cellulose dyeing, theory of dyeing of silk & wool was focused initially adsorption of the dye to the fibre surface from the dyeing bath, secondly diffusion of the dye into the interior of the silk fibre and finally fixation of the dye with reaction centers in fibroin [13]. At last but not least, changing the important dyeing parameters like fixation time, dyeing temperature, dye bath concentration, pH of the solution etc of dyeing equilibrium to develop a unique dyeing process of silk & wool with reactive dyes [14].

6. EXPERIMENTAL

6.1. Materials

All the used raw materials were of commercial grade and used without further purification. Silk sample was obtained from Hangzhou Dali Silk Co., Ltd. Sodium Carbonate (Na_2CO_3), Sodium sulphate (Na_2SO_4), Acetic Acid (CH_3COOH) & Detergent are used from lab.

LIYUANSOL FL: These fluorine based reactive dyes were obtained from Color Root (Hubei) Technology Company Ltd., China. Liyuansol series is a fluorine based reactive dyes contain a vinyl sulphone and a triazine group, they are economic and often used in the dyeing and printing process. Full range of shades and have some brilliant dyes. Those series are widely used for cotton dyeing.

LANASOL: These reactive dyes were obtained from Huntsman Dyes. The LANASOL® range consists of sulfo-group-containing reactive dyes which have been especially developed for wool dyeing. They contain 1 or 2 bromo-acrylamide reactive groups which form a covalent bond with the nucleophilic groups of the wool's amino acids during the dyeing process, resulting in outstanding wet fastness properties.

6.2. Methods

Here dyeing process was carried out by the common process of reactive in a bath having material to liquor ratio of 1:50 with different condition of the dyeing equilibrium in an automatic laboratory dyeing machine. Of course the dye manufacturers recommended processes were also followed initially. The dye solution was prepared initially by adding dye (owf) and distilled water at room temperature. Wet fabric was added to dye-bath then the dyeing process was inaugurated after 5 minutes. Raising the temperature at 80C & continue up to 25 minutes. The temperature was raised again to 90C temperature and dyeing was carried out 95 min. The pH of dyeing solution was maintained with acetic acid and sodium carbonate. The dyed material was then rinsed at hot (90C) water for 10min followed by cold water to remove the unfixed dye and dried at room temperature.

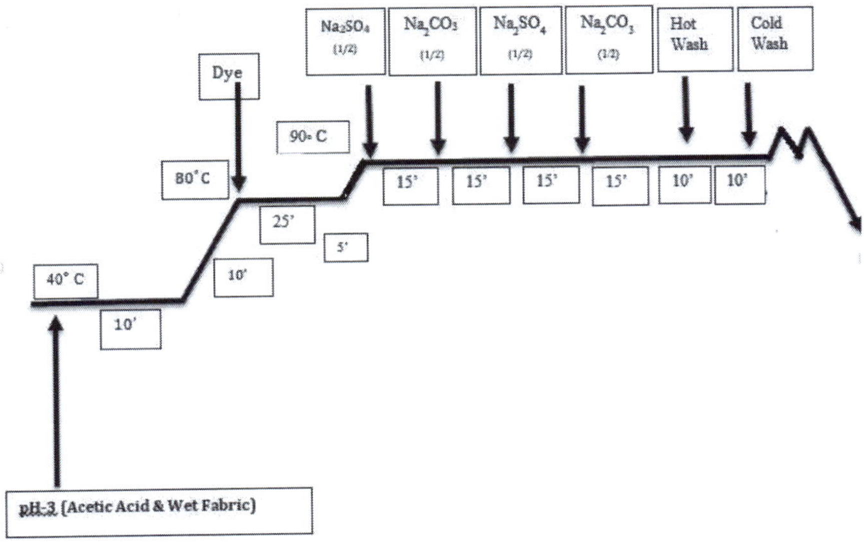

Figure 2. Process curve of silk dyeing with reactive dyes

6.2.1. Determination of K/S Value

K/S value of the dyed fabric is measured by using spectrophotometer (DATACOLOR®-110).

6.2.2. Determination of Dye Uptake%

Dye concentration was measured by a UV/VIS spectrophotometer and calculated dye uptake according to following equation:

$$Up\ Take\ \% = \frac{C_0 - C}{C_0} X\ 100\%$$

C_0 → Concentration of the Dye in solution before dyeing

C → Concentration of the Dye in solution after dyeing (solution of after dyeing + solution after hot wash).

For the accuracy of the result above equation was applied both dyeing & washing step. Finally resultant dye take up % was calculated.

7. RESULT & DISCUSSION

7.1. Evaluation of Silk Dyeing with Reactive Dyes

Here dyeing performance is assessed by the time, appropriate pH, dye bath concentration, fixing time, alkali dosage, amount of salt dosage as well as promoter consumption to calculate an efficient dyeing principle by altering the important variable parameters on dyeing equilibrium.

7.1.1. Effect of pH on Silk Dyeing with Reactive Dyes on Dye Uptake and K/S Value

During observe the dyeing result under different pH, the other conditions were as follows-

- Material and liquor ratio is 1:50
- Dye is 3% (owf)
- Na_2SO_4: Na_2CO_3 = 1:1
- Time 100-120 min and
- Temperature 80-90C.

pH has a great influence on silk dyeing. Silk containing $-NH_2$ groups are protonated ($-NH_2^+$) in the acidic solution. The more amounts of $-NH_2^+$ groups mean more chemical bonds with dyes. The Figure 4 shows that for both types of reactive dyes the dye uptake % is decreasing with the increasing of pH value of the dye bath. The dye uptake for both Lanasol (81.80%) and Liyuansol (67.00%) are highest at pH 3. Although at pH 4 the uptake of dye for Liyuansol is very good (62.03%) but for Lanasol has dramatically decreased (38.26%).

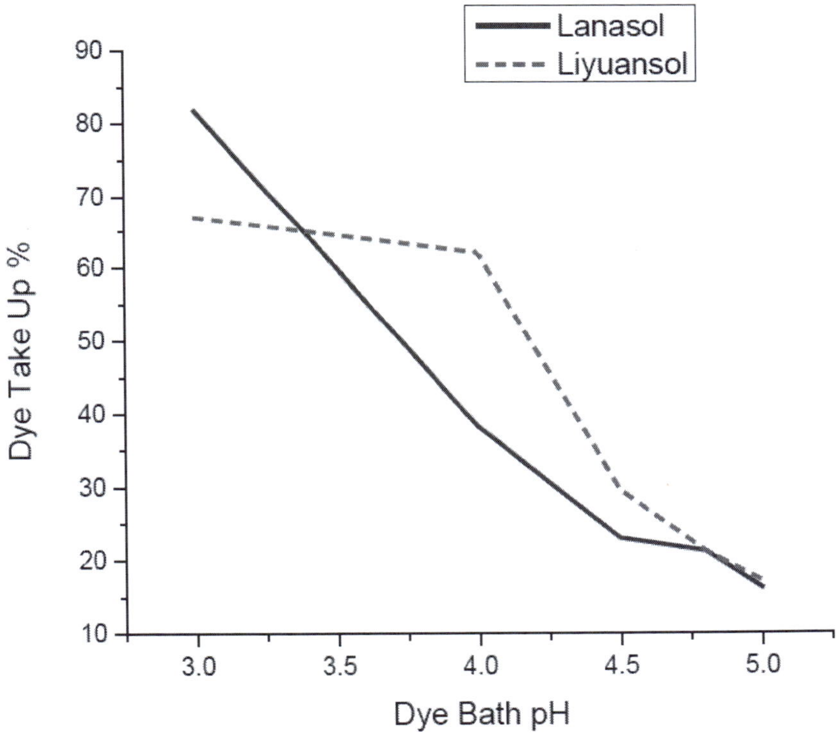

Figure 3. Effect of pH on dye up % for Lanasol & Liyuansol Dyes

As well as dye uptake the K/S of silk dyeing shows the similar result under different pH. K/S value is very high for both Lanasol and Liyuansol at very low pH; 3. When the pH increased at 4, the K/S of Lanasol decreased directly (6.95) and the K/S of the Liyuansol is almost stable (11.27). But for both dyes the K/S has decreased with the increase of the dye bath pH.

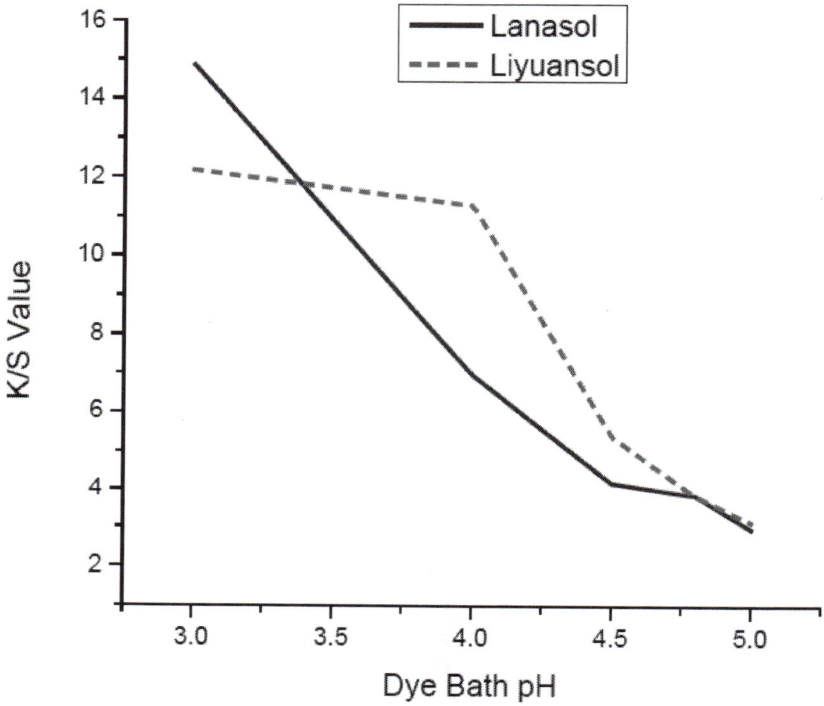

Figure 4. Effect of pH on K/S value for Lanasol & Liyuansol dyes

From Figure 3 & Figure 4 it is clear that the dye uptake and K/S both are very high at pH 3. And for Lanasol the results are changed noticeably while the results are stable for Liyuansol pH between 3-4.

7.1.2. Effect of Time on Dye Uptake and K/S Value during Silk Dyeing

The parameters were remaining unchanged except (see section 7.1) time (during the observation under different pH).

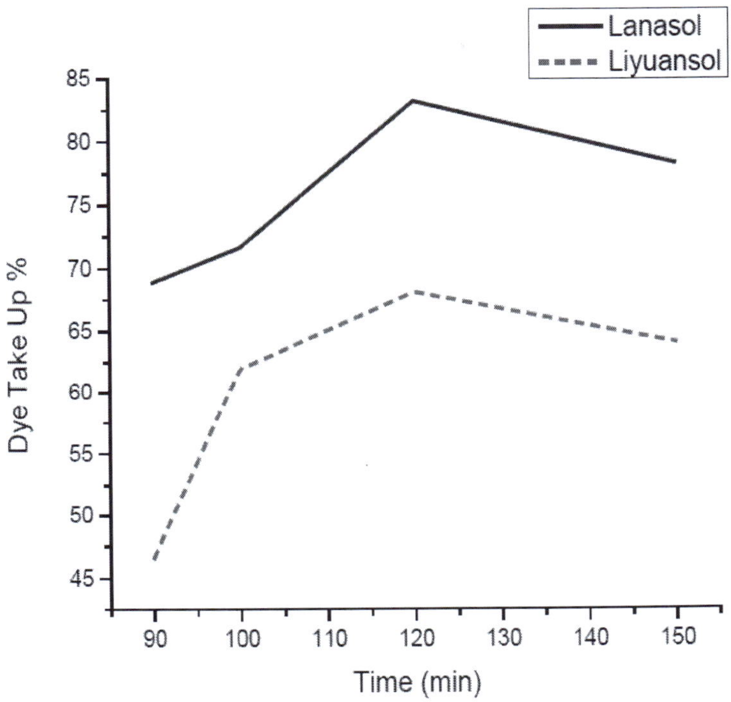

Figure 5. Effect of dyeing time on dye uptake

The optimum dye uptake and K/S value were found at dyeing time 120 min. (2hrs) at constant temperature 90C. Below 120 min and above 120 min the results were unsatisfactory. When the time goes more than 120 min the dye uptake decreased due to hydrolysis of some of the dyes.

The dye uptakes for Lanasol and Liyuansol at 120C are respectively 83.26% & 68.15%; and the K/S values are 17.22 & 12.38 respectively.

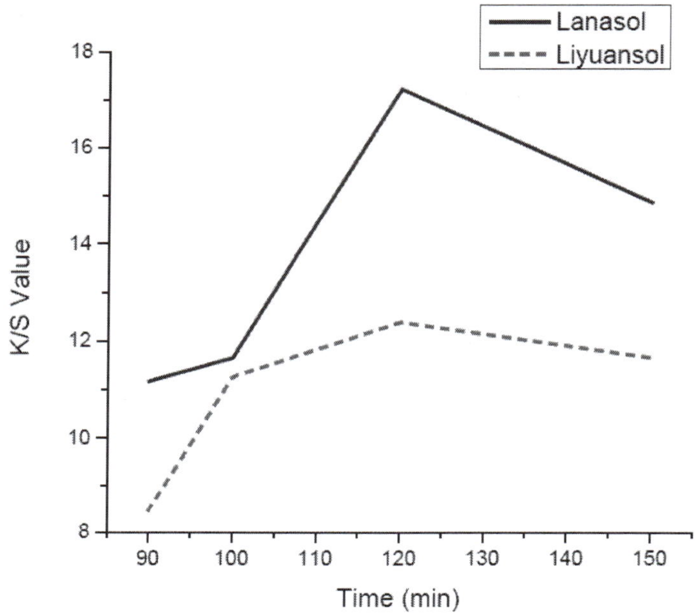

Figure 6. Effect of dyeing time on K/S value

7.1.3. Effect of Na_2SO_4 & Na_2CO_3 on Dye Uptake & K/S Value of Silk Dyeing

Firstly the effect of salt is observed by dosing Na_2SO_4 & Na_2CO_3 after the dosing the dyes and acetic acid, and then dyeing is finished by above-mentioned method. But it was found huge color spot and uneven shade on the fabric.

The effect of Na_2SO_4 & Na_2CO_3 are observed under the dyeing condition as follows- Material and liquor ratio is 1:50, dye is 3% (owf), pH 3.5, time 100-120 min and temperature 80-90C. Na_2SO_4 & Na_2CO_3 both salts are used during dyeing. The ratio of Na_2SO_4 & Na_2CO_3 is changed like 1:4, 2:3, 3:4, 4:1 and 1:1. The results for both dye uptake and K/S value are recorded in Table 1.

The results were good at 1:1 for Lanasol. Although the results for Liyuansol was very good at ratio 3:4 but, to reduce the salt consumption the ratio of Na_2SO_4 & Na_2CO_3 is used 1:1.

Table 1. Effect of Na2SO4 & Na2CO3 on dye uptake & K/S

Salt Ratio	Salts		Lanasol		Liyuansol	
Na_2SO_4 : Na_2CO_3	Na_2SO_4 (g/L)	Na_2CO_3 (g/L)	Dye Uptake (%)	K/S Value	Dye Uptake (%)	K/S Value
1:4	5	20	78.11	11.07	54.27	9.86
2:3	10	15	56.22	8.15	47.39	8.61
3:4	15	20	67.48	13.55	55.37	10.06
4:1	20	5	49.44	10.23	38.58	7.01
1:1	10	10	83.24	17.22	51.19	9.30

8. APPLICATION PERFORMANCE TESTING

After getting of optimum dyeing condition above dyeing process evaluation, silk fabric was dyed in different dye concentration (1%, 3% & 5% owf) by using two commercial reactive dyes. Their different application performance tests were tested and compared with acid dyed silk sample that was dyed followed by general acid dying procedure.

8.1. Effect of Dye Concentration on Dye Uptake

Dye uptake enlarged gradually as dye concentration increased. With cumulative concentration, further dye moved to fabric and the deepness of color became stronger. From the above column illustration it is clear that the reactive dyes exhibit advanced dye uptake than acid dyes at higher colorant concentrations.

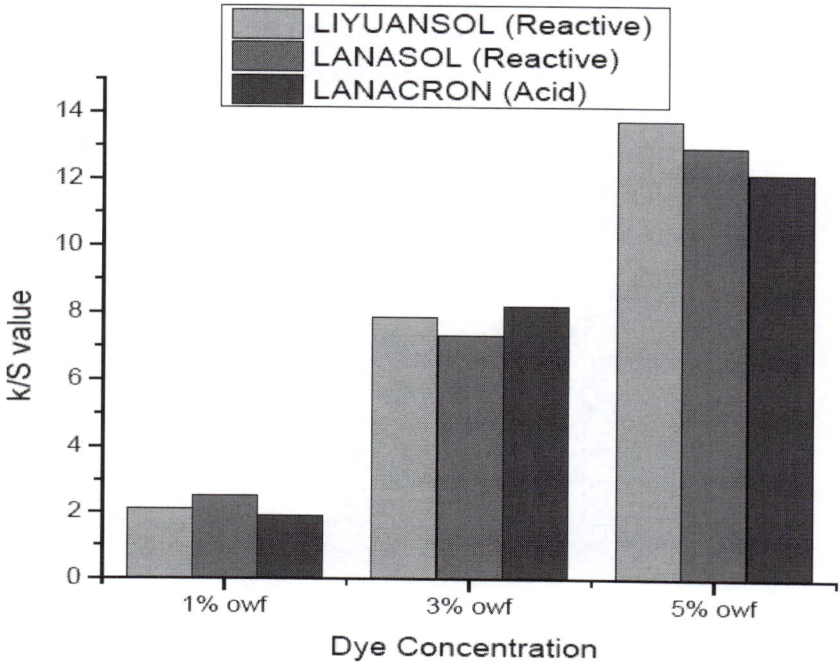

Figure 7. Effect of dye concentration on K/S value

8.2. Color Fastness to Wash

Dyed sample is prepared with sized 10 x 4 cm is taken and is sandwiched between two adjacent fabrics and stitched. The sample and the adjacent fabric were washed together. The washing solution was preheated to the required temperature of washing. The liquor ratio 1:50. After soaping treatment, the specimen is removed, rinse twice in cold water and then in cold running tap water. Squeezed and dried in air at a temperature not exceeding 60°C. The value is evaluated with the help of grey scales.

Figure 8 Shows that CF to wash results was excellent for all concentration above 4. Only for Lanacron 5% shade the ratting is 4.

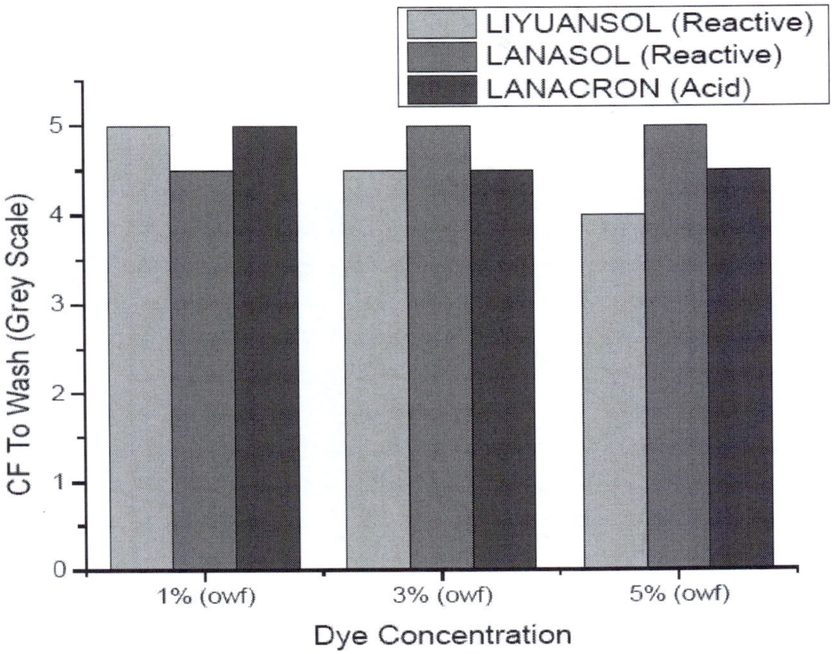

Figure 8. CF to Wash for different Dye concentration

8.3. Color Fastness to Dry-Rubbing

The test is quite sensitive and for getting consistent result, it is necessary to use standard crockmeter cloth, maintain uniform pressure for applying rubbing strokes and number of strokes. Degree of staining is visually assessed using Grey scale for change of color.

The test showed excellent results on CF to dry rub. Figure 9 shows that, the result of CF to dry rub for both dyes (Lanasol, Liyuansol & Lanacron) excellent (4-5 & 5) for all dye concentration.

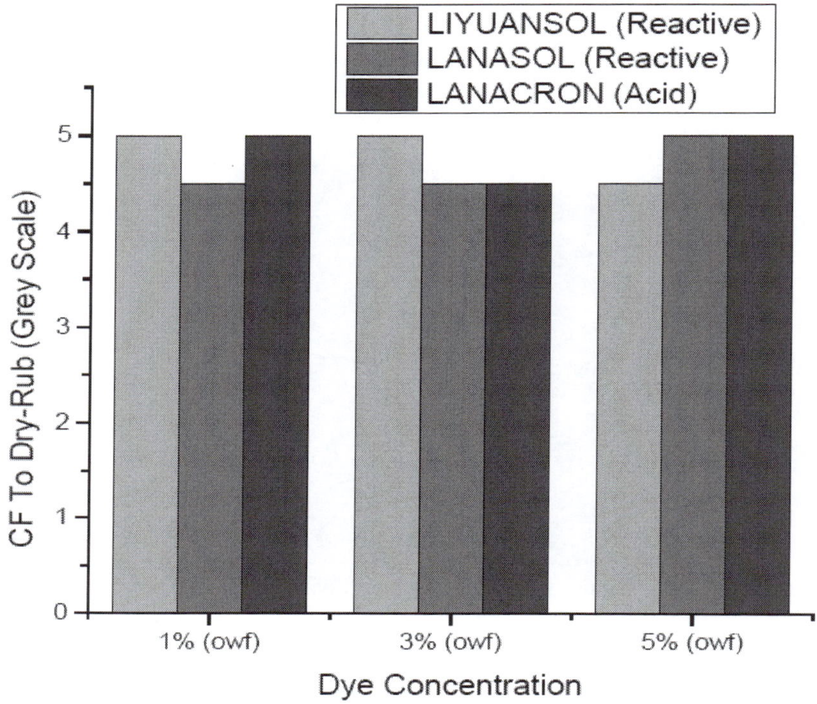

Figure 9. CF to Dry-Rub for different dye concentration

8.4. Color Fastness to Wet-Rubbing

In the case of wet rubbing, %moisture on the crock-cloth has to be kept to uniform level and rubbing cloth that has been wetted with water, has to be squeezed to contain its own weight of water. For AATCC 116-1995 methods, wet pick up is to be maintained between 65 ±5% by squeezing the wet crockmeter cloth using a AATCC blotting paper. Any variation in the moisture content can lead to deviation in the rating. With high amount of moisture i.e.., wet pick up, ratings will be lower.

The Figure 10 shows the CF to wet rub for 5% dye concentration is a bit lower than the 1% & 3% but the overall results shows very good results.

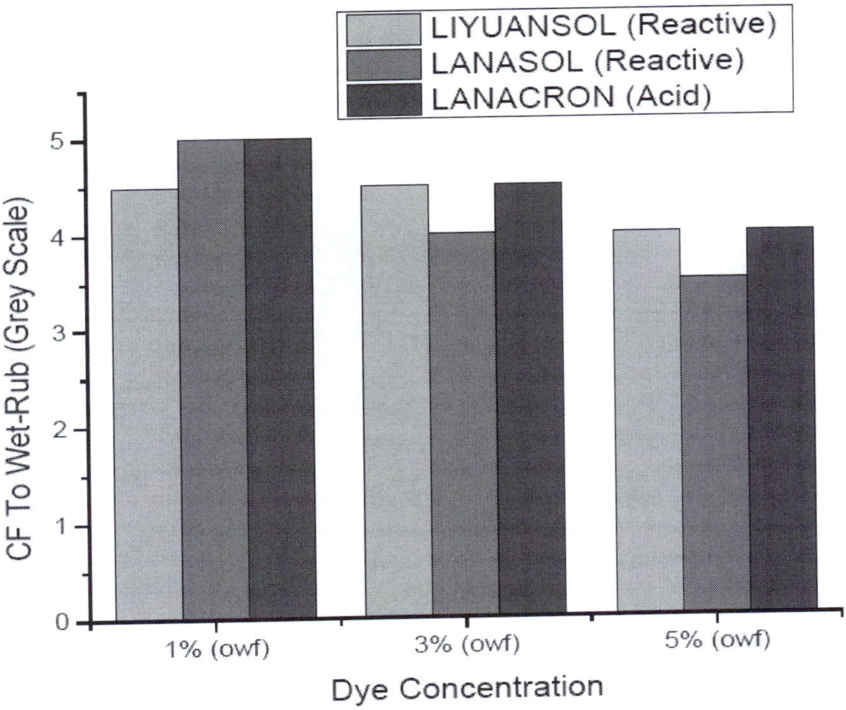

Figure 10. CF to Wet-Rub for different dye concentration

8.5. Color Fastness to Perspiration

The fastness of dyed fabric with respect to acidic & alkaline perspiration was evaluated. For the alkaline (pH-8) and acidic (pH-5.5) liquors were prepared and the composite specimens were dipped in acidic and alkaline solution separately for 30 minutes. After ensuring of good and uniform penetration solution, the liquor was poured off and the excess water and air bubbles, if any were removed by passing the specimens in between two glass rods. Composite specimens were then placed between glass/acrylic plates with a pressure of 12 kpa perspirometer. The perspirometer was kept for four hours at a temperature of 37 (±2C) Afterwards, the fabrics were removed, separated and dried in air below 60°C. The values were rated as per the grey scale. The assessment of colorfastness to perspiration for applied dyes and result change according to the greys scale reading for color change are shown in the Table 2.

Table 2. Color fastness to perspiration (Result for color change)

Applied Dyes	Solution Nature	Grey Sale Value (color change)		
		1% Dye Conc.	3% Dye Conc.	5% Dye Conc.
Liyuansol (Reactive)	Acidic Solution	5	4 - 5	5
	Alkaline Solution	4 – 5	5	4 - 5
Lanasol (reactive)	Acidic Solution	4 – 5	5	5
	Alkaline Solution	5	4 - 5	4 - 5
Lanacron (Acid)	Acidic Solution	4 – 5	5	4 - 5
	Alkaline Solution	5	4 – 5	5

From the assessment of Table 2, it can be said that the color fastness to perspiration for the color change according to grey scale for applied dyes both specially manufactured reactive dyes for protein fibre (Lanasol), commercial Fl based cotton-reactive dyes (liyuansol) and available acid dyes (lanacron) for silk/wool are shows satisfactory result.

9. CONCLUSIONS

This research shows that the specialized reactive dyes (Lanasol) specially designed for protein fiber (wool) dyeing exhibit good properties on silk. Also another Reactive dye (Liyuansol FL); that is fluorine based which is specially manufactured for cotton dyeing is suitable for silk dyeing. The dyeing procedure and auxiliaries were same for both dyes. So, fluorine based reactive dyes (Liyuansol Fl) for cotton can be used for dyeing silk fiber.

The quality test result of silk with Fluorine based reactive dye shows very good to excellent result as the specialized wool dye (Lanasol) shows. It is also found an excellent or equal application performance as the acid dyes (Lanacron) which specially used protein fibre dyeing. The dyeing temperature was below boil so; the less or no degradation of protein as well as less dye migration will be happened.

So, it can be recommend to use fluorine based reactive dyes for silk dyeing in textile industry to overcome the quality as well other processing problem that occurs with during protein fiber dyeing with Acid dye. Also it will be cost effective in term of using less auxiliaries, low energy to boil, less finishing chemicals to improve quality of acid dyed fabric.

The homogeneity of the dye on various textile products in real baths needs further studies.

ACKNOWLEDGEMENTS

We would like to express our gratitude to Professor Jiang Huiyu (Vice Professor, School of Chemistry & Chemical Engineering, Wuhan Textile University) for his inspiring guidance, encouraging attitude and valuable suggestions throughout this research work. We would also like to thanks Mr. Chen for his co-operation during practical experiment.

We also want to mention the mane of Color Root (Hubei) Technology Company Ltd.; because this work was financially supported by this company.

REFERENCES

1. Shin, Y., Dyeing Properties and Color of Silk Fabrics Dyed with Safflower Yellow Dye. Journal of the Korean Society of Clothing and Textiles, 2008. 32.

2. Liu, J. and H. Wang, Union Color of One Bath Dyeing to Silk/Flax Fabrics with reactives. Advanced Materials Reserch, 2011. 332-334.

3. Zhang, G., L. Lu, et al., Dyeing Acceleration Theory of Molecule Companion on Silk with Reactie Dyes.Advanced Materials Research, 2102. 549.

4. Broadbent, A.D., Basic Principles of Textile Coloration. 2001, England: Society of Dyers and Colourists

5. Rush, J.L., Dyeing Primer. AATCC Journoul. 1981. 7-9.

6. Lewis, D., Damage in wool dyeing. Review of Progress in Coloration and Related Topics. Society of Dyers and Colourists, 2008. 19(49–56).

7. Silk fibre, Physical and Chemical Properties of Silk Fibre, . June 27, 2012 [cited 2015 April 05, 2015]; Available from: http://textilefashionstudy.com/silk-fiber-physical-and-chemical-properties-of-silk/.

8. Bae, J.S., J.H. Park, et al., Dyeing and Fastness Properties of a Reactive Disperse Dye on PET, Nylon, Silk and N/P Fabrics. Fibers and Polymers, 2006. 17.

9. Uddin, M.K. and M.S. Hossain, A Comparative Study on Silk Dyeing with Acid Dye and Reactive Dye.International Journal of Engineering & Technology IJET- IJENS 10(6).

10. Protein fiber, the silk polymer and its system, NPTEL, E-learning courses from the IITS and IISC.

11. LEWIS, D.M., Handbook of textile and industrial dyeing. Vol. 1. 2011, UK: Woodhead Publishing Limited.

12. How dyes attach themselves to fabric, Salter's Chemistry. April 05, 2015]; Available from:http://www. 4college. co.uk/a/Cd/fabric.php.

13. Koussoulou, T., Photodegradation and photostabilization of historic silks in the museum environment.Institute of Archaeology UCL, 1999.

14. Hossain, M.F., Practice of textile coloration. Vol. 1. 2007, Dhaka: Book Fair Publications.

10

CHAPTER

Progress on Fabric Electrodes Used in ECG Signals Monitoring

Zhen Liu, Xiaoxia Liu

College of Fashion, Shanghai University of Engineering Science, Shanghai, China

ABSTRACT

Wearable monitoring system is designed for skin stimulation of conductive adhesive, prolonged physiological monitoring and biocompatibility, whose core is fabric electrodes and it can feedback physiological status by analysis of abnormal electrocardiogram (ECG). Fabric electrode is a sensor to collect biological signals based on textile materials including signals acquisition, processing systems and information feedback platform and so on. In this paper, the design methods and classification of medical electrodes would be introduced. It also sorted out the principle of bio- logical electrical signals, the design methods and characteristics of different material and different structure electrodes from the point of dry electrodes and wet electrodes. There are many methods that can be used to prepare fabric electrodes. They are mainly metal plating, conductive polymer coating, magnetron sputtering, gas phase deposition and impregnation. Besides, they select the appropriate substrate, conductive medium and composite way to get light fabric electrodes which have high conductivity, good conformability. From the perspective of biological signal acquisition by fabric electrodes, this paper also sorted out the influence and approaches of biological signals and the way to feedback the physiological condition of human. As a new generation of bio-signal acquisition material, fabric electrode has met the requirements of the development of modern medicine. Fabric electrode is different from traditional conductive materials in the characteristics of comfort, intelligence, convenience, accuracy and so on.

Keywords: Fabric Electrode, Biological Signals, Slide Artifacts, ECG

1. DEVELOPMENT OF FABRIC ELECTRODES

Textile science is no longer a discipline just for clothing. With the accelerated pace of development of textile science and medical technology, more and more attention has been paid to medical textile and industrial textile research. Recently, the quality of life was seriously affected by cardiovascular disease, heart disease, chronic pneumonia and other diseases, and more researches about physiological condition monitoring were followed. A long-term monitoring system for old people's physiological state came into being, and health monitoring systems typically included signal acquisition, processing systems, signal contras and physiological state feedback platform. Textile materials have developed in electronic information technology very well which makes it the best choice to use intelligent fabric electrodes as sensors for monitoring system due to intelligence and flexibility. As a new generation of bio-signals acquisition sensor, fabric electrode met the requirements of modern materials like convenience, intelligence, accuracy and so on. It is different from traditional needle electrode which is prone to cause skin damage and has some other shortcomings. Fabric electrodes belong to the category of smart textiles in the electronic textile which is not just a simple conductive fabric. The smart fabric electrode consists of micro power, connecting member and signal path together to monitor the biological electrical signals. Fabric electrode is provided with high research value and social economic benefits by the great market demand in medical field.

Electrode is used as a sensor to collect biological signals which begin at an embedded motherboard. Georgia Institute of Technology embedded GTWM motherboard into clothing to detect life signs of wounded soldiers in 1996. GTWM wearable motherboard made of optical fibers concentrated human life, biological signals monitoring and signals processing functions into one. How to collect biological signals by wearable monitoring system and feedback accurate physiological state has been a hot topic in foreign countries in recent years. Researches have paid a great attention to the research of fabric electrodes. As shown in Table 1, MagIC project [1] of Polytechnic University of Milan, Italy used fabric electrode as an original information collection to monitor heart and lung function by acquiring biological electrical signals of breathing and temperature changes. Jacket monitoring system prepared by the EU WEALTHY project [2] used textile conductive wire, Bluetooth and GPRS technology to transmit and get feedback on dynamic bioelectric signals in mechanical movement. In addi- tion to these projects, MyHeart project [3] of monitoring the elderly cardiovascular disease of Philips Research Europe, Vital Jacket plan of the University of Aveiro [4] and CHRONIOUS program [5] of Bremen University had also used a similar embedded technology to collect biological signals through embedding fabric electrodes in wearable clothes and transmitting signals by wires or wireless technology. The research of wearable body monitoring system has made great achievements in the monitoring of elderly breathing, body temperature and blood oxygen saturation. In addition, monitoring time of human physiological signals is short and further research is required.

2. GENERATION PRINCIPLES AND CHARACTERISTICS OF FABRIC ELECTRODES AND BIOLOGICAL SIGNALS

2.1. Characteristics of Fabric Electrodes

Body weak electric current is called biological electric current which results from heart beats, muscle contraction and other actions. Human body composed of cell membrane is a kind of semi-permeable membrane. The principle of cell membrane is that only a certain kind of ions can go through it in a certain state which results in a potential difference between inside and outside of cell membrane. Fabric electrode could be essentially regarded as an interface because of achieving ions transformed between skin surface and wires to analyze electrical current. Biological signals passed through tissues and fluids surrounding the heart to the conductive skin surface and formed a weak potential difference. Electromyography signal is a kind of weak physiological signal caused by mechanical heart beat. The cell membrane of human body is similar to a semi-permeable membrane.

2.2. Generation Principles and Characteristics of Biological Signals

The principle of semi-permeable membrane is that only a certain kind of ions can go through it in a certain state. It's known that there are a series of physiological and chemical changes in the body when a man is in motion. At different potentials, the cell membranes have a different permeability for sodium ions (Na^+) and potassium (K^+). That is to say, when the cell is at resting potential, membrane permeability for potassium (K^+) is greater than sodium ions (Na^+), making potassium ions move to the extracellular, and the membrane potential outside is higher than the membrane potential inside, forming a negative potential inside but positive outside. Instead, when the

Table 1. Research status of fabric electrodes.

Project title	Basic component	Communication mode	Measurement object	Medical applications
MagIC	Vest with sensors embedded/portable circuit board	Bluetooth technology	ECG/breathing/ temperature	Record the signal of cardiopulmonary function and daily activities
WEALTHY	Vest with sensors embedded	Conductive yarn/ Bluetooth technology/GPRS	ECG /breathing/ Temperature/EMG signals/ mechanical activity	Monitoring the recovery of elderly chronic disease
MyHeart	Chest belt with sensors embedded/PDA	Conductive yarn/ Bluetooth technology/ GPRS	ECG/other important signals/mechanical activity	Prevention of cardiovascular disease in elderly
Vital Jacket	Vest with sensors embedded/PDA/PC	Conductive yarn/ Bluetooth technology/ GPRS	ECG/breathing/temperature/ mechanical activity /oxygen saturation	The elderly physiological signs monitoring
CHRONIOUS	T-shirt with sensors embedded/PDA/PC	Conductive yarn/ Bluetooth technology/ GPRS	ECG/other important signals/mechanical activity	Elderly chronic disease prevention

cell is at action potential, membrane permeability for sodium ions (Na^+) is greater than potassium (K^+), making sodium ions (Na^+) continuously transfer to the intracellular, and the membrane potential outside is lower than the membrane potential inside, forming a negative potential outside but positive inside [6] . Biological signals have some characteristics as followed.

1) Signal is weak. Biological electrical signals are generated by the difference of random ion concentration inside and outside of human cell. The signal collected must be gone through a relative larger common-mode rejection and higher amplification circuit. In addition, biological signals can be affected by motion artifact and noise and it also needs to be amplified step by step. Electricity and magnetism are inseparable. Bioelectrical signals captured by fabric electrode are susceptible to being affected by current source, especially the electromagnetic waves around 50 Hz frequency. Biological signal in power frequency noise is random and it could easily drown out the pure electric signal. Therefore, electromagnetic frequency closed system reducing the wire length method has to be used to reduce the power frequency noise to acquire relatively pure signal. Due to the influence of external electric field and electromagnetic field, there is some potential on the body surface of patients. This kind of induced potential is bonded to passing through electrodes to the preamplifier to form interference signals. It usually presents 50 Hz frequency interference which can be controlled by increasing the amplifier's common-mode rejection ratio in laboratory conditions.

2) Noise processing. Because the skin friction noise and electromagnetic impact of external environment can't be ignored, biological signals collected must be passed through low pass and high pass filtering process to gather accurate electromyography signals. On the other side, baseline drift caused by human respiration and the circuit connection between fabric electrode and skin would make a great influence on electrical biological signals. Baseline drift could make the band offset the original position and affect the accuracy of low frequency ECG signals. Therefore, in order to obtain a certain output signal, the treatment of electrical signals through amplifier has very strict requirements.

3) Large difference in signals. The difference of bioelectricity is great. As we know, people of different ages have different metabolisms.

The ion concentrations and temperature are not equal inside and outside of cell membrane. Studies have shown that human skin has some small potential difference, this potential difference is related to human mechanical movement. When the fabric electrode collecting biological signals, the accuracy would be disorganized and affected by skin surface. On the other side, human mechanical movement makes the skin extrusion and stretching deformation, but the fabric electrode could not be completely deformed with the skin and sliding artifact could be caused by skin-electrode interface due to the relative sliding friction. Thus, the biological signal is very complex which needs high-pass filter technology, low-pass filter technology, 50 Hz notch filter and other professional signal processing technology [7] - [9] .

3. DIFFERENT KINDS OF ELECTRODES

Impedance and frequency characteristics of human epidermal tissue should be taken into consideration before studying human physiological state through collecting biological signals. The level of the potential difference and impedance is one of the most important factors to determine the selection of electrodes. From a biological point of view, human skin can be regarded as a pure resistance for its good conductivity. The cuticle thin insulating film acts as a medium between the two electrode plates of capacity consisting of dermal tissue and electrode plate which determines the skin impedance magnitude. The sweat caused by longtime mechanical activity would increase the degree of fit between human skin and fabric electrode and decrease the impedance caused by gap air between the electrode and skin. Fabric electrode can not seamlessly fit the skin. The gap air between the electrode and skin would also affect the conductivity which can be seen as an equivalent circuit of resistors and capacitors in parallel [10] [11] .

As shown in Figure 1, electrodes can be divided into insertion electrodes and surface electrodes and surface electrodes can be divided into dry electrodes and wet electrodes. At the state of rest, there are a certain number of positive ions inside of cardiac muscle cell and the same amount of negative ions outside. When the cardiac muscle cells get a certain extent of stimulus, the permeability of cardiac muscle cell would change and the positive ions would rush into or exhaust from the cell membrane and form an electric potential difference. The electric potential difference is the foundation of electrode collecting biological signals. Needle electrode is a typical insertion electrode, which collects every moving part signal by inserting it into the human skin. Although the signal acquisition method can accurately collect biological signals, it needs not only high operation requirements, but also can cause severe harm to human body.

3.1. The Characteristics of Dry Electrodes

The study of dry electrode has caused more and more attention in recent years. Dry electrode is different from wet electrode, which needn't apply conductive adhesive to improve conductivity but it can also cause great motion artifacts and noise by skin friction. Biological signals acquired by dry electrode must be filtered by buffer amplifier circuit and transformed from high impedance input into low impedance input to reduce interference noise and motion artifact. In a word, the design of fabric electrode needs considering every step of process including selection of materials, structure of fabric and conductive way of biological signals. Fabric electrode could be easy to fix in clothes or elastic bands to make patients comfortable. This kind of flexible electrode has the advantages of breathability, moisture permeability, soft and biological compatibility. Hongyi Zhai [12] et al. use wearable fabric electrodes made of silver-plated fabric to acquire ECG to combine it with classic ECG after signal conditioning and micro-processing device and get abnormality ECG eventually.

3.1.1. The Affects of Different Fabric Structure

Fabric organizational structure not only affected the firm level between conductive materials and fabric-base materials but also the adhesion of conductive materials. At the same time, structure can also make some influences on the stability of signal acquisition and the comfort of wear clothing. Besides the size and the stability of fabric electrodes, fabric organizational structure directly affects the degree of fitting between the electrode and skin, especially the side contacting with human. Almost all fabric organizational structures have been selected as fabric electrode materials like weaving, knitting, embroidery, jacquard fabric and non-woven fabric, but it remains many problems about structure which needs future researches. 1) Woven fabric electrode. Yarns should be weaved into plain woven fabric on a loom first, then plated with a metal surface on the sample, it would be very hard for us to guarantee good surface uniformity of conductive metal and surface morphology everywhere. To some extent, requirements of metallization technology are very strict because the degree of surface uniformity of conductive metal would affect the degree of chlorination irregularity directly. Dandan Tao [13] et al.

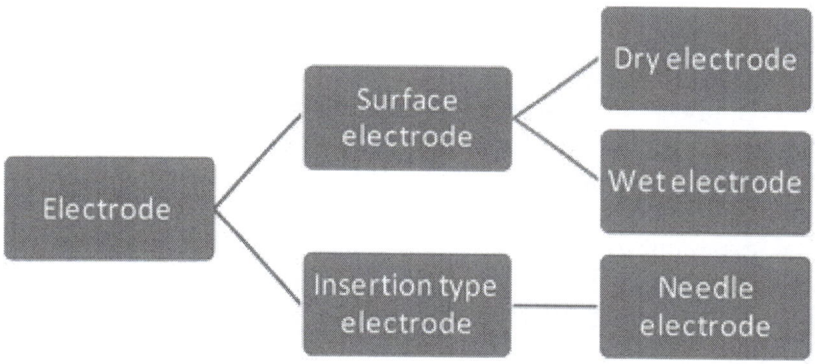

Figure 1. Classification of electrodes.

made conductive silver paste and fabric into an woven fabric electrode which is fixed by elastic belt, and used as a collecting element of EMG signals during mechanical movement. On one hand, warps and wefts consist of woven fabric together and the knot points of fabric can be seen as resistors when the woven fabric is used as a fabric electrode. On the other hand, some capacitive characters would be produced between the space portions of fabric electrode. In a word, plain woven fabric could not only be regarded as a sample resistor or a capacitor, but it's also an electronic component consisting of resistors and capacitors. With the impact of common capa- citors and resistors together, the woven fabric electrode would receive different quality signals in different fre- quencies or different pressures. 2) Knitted fabric electrode. Silver plated yarns could be usually chosen as basic materials manufactured into a three-dimensional cylindrical knitted fabric electrode on a hosiery knitter then cut into a desired shape sample. Coil structure determined an unstable fabric structure and was easy to deform, so the program of conductive metallization is unfeasible for knitted fabric electrode. From the point of service life, unstable dimensional and discontinuous conductive metal is the reason that knitted fabric electrode is not as good as woven fabric electrode. However, knitted fabric electrode can fit human skin better and deform with it. Finally, this feature of good elasticity could reduce the noise impact caused by relative friction between skin and electrode [14] .

These structures have many advantages and disadvantages. First, not only woven fabrics have a stable warp and weft structure but the production process is very complex. Secondly, knitting fabric is more convenient than woven fabric but the size of coil structure is unstable because the yarn is easy to slide. Similarly, embroidery fabric is very convenient to design because of the unique structure. Non-woven fabric electrode has the same characteristic but the structure is not as stable as embroidery fabric for its unique structure between fibers. All in all, weaving, knitting and embroidery organizational structure are all popular organizational structures for fabric electrode.

3.1.2. Properties of Different Fabric Electrodes

1) Metal fabric electrodes. Metal fabric electrodes can be designed in the way of disposing conductive metal inside the fabric or yarn surface for transferring electrical signals. It is not suitable for fabric electrode because the metal thread couldn't meet the requirements of washable and anti-wrinkle. Recently, metal conductive fibers stretched from metal yarns are usually used to blend with common yarns, which can not only solve the problems of metal threads but it can also output more simple electrical signals. Metal conductive yarns have good conductivity, heat resistance and also have certain strength. In view of the above features, metal conductive is more suitable for ECG electrodes to transform the ion current on skin surface into an electron current in wire. It also has some shortcomings, although such an electrode retains the degree of fabric softness, the metal fibers can result in skin tingling easily if the implantation skill is not good. Metal materials can also be used as a dense conductive metal layer on the surface of fabric by magnetron

sputtering or vapor deposition methods. Metalized fiber blended fabric electrode depends on the principle of "silver mirror reaction" to form a layer of silver on the surface of fibers by redox reactions and blend with other fibers to weave into fabric electrodes. In addition to the electrodes for ECG, such conductive fabrics can also be used as radiation materials for the excellent electromagnetic radiation protection. Binbin Yue and Xin Ding [15] plate gold film on fabrics by magnetron sputtering, then deal with acetonitrile solution, which containing toluene sulfonic acid and polypyrrole monomer to obtained the polypyrrole composite conductive fabric electrodes by chemical polymerization method polymerization methods. Fabric electrode prepared by this process has the characteristics of a uniform coating, strong adhesion, washable and so on.

2) Poly fabric electrode. Polymer fabric electrode. Xiaolan Fu [16] et al. from Donghua University selected polyaniline and metal composites materials as an intermediate flexible conductive layer. Under ultrasound conditions, they use polyester as base materials to prepare polyaniline polymer/Cu composite conductive fabric with a chemical coating method. The reasons why conductive poly can be used for fabric electrodes can be explained from three aspects, namely field emission effect, percolation and tunneling effect [17] . 1) Field emission effect. According to the field emission effect, as long as ions of different potential with a certain charge close enough, electrons can leap over the intermediate insulating layer to the adjacent ion and generate a current, then non- contact-conductive is achieved. 2) Percolation. Based on percolation, as long as the concentration of charged particles in a conductive polymer is high enough and the distance between conductive particles is in nanometer range, the electrons can form current. 3) Tunneling effect. Tunnel theory holds that when the thickness of isolation layer between ions is less than 10nm, an electron tunnel could be formed by electrons to produce electron flow. These three theories supported the reason why polymer can be used to prepare conductive layer fabric electrodes.

3.2. The Characteristics of Wet Electrodes

Wet electrode is a kind of electrode with conductive adhesive. Conductive adhesive is made of high concentration negative ions electrolyte. These materials can improve the electrical conductivity of the skin surface and reduce the error of biological signals but it may cause skin irritation like skin allergy or itchy in prolonged exposure because the chemical properties of adhesive would damage human skin. Electrode with conductive adhesive is not suitable for long time monitor as conductive adhesive would volatilize in the process of use and affect the signal quality. As the most common bioelectricity measuring wet electrode in clinical medicine, metal plate electrode connects human skin and induces surface bio-potential with metal conductive materials. The structure of metal plate electrode is very simple. A layer of conductive adhesive or sponge mat impregnated with electrolyte should be used to pad the vacancy to improve the contact area between the skin and electrode. In some cases, electrolyte conductive materials were used to improve the conductive ability and surface smooth performance. Wet electrode conductive adhesive is usually attached with an electrochemically-treated silver chloride coating on the side of contacting with the surface of skin. There is a lead wire welded on the back of metal plat electrode and insulation materials such as polyvinyl chloride and epoxy resin used as connecting materials to protect lead wires and the electrode. This kind of electrode was usually used to record electrocardiogram signals on the patient's chest with the layer of electrolyte in shot time. Besides electrocardiogram (EEG), wet electrode can also record electromyography signals (EMG) and electroencephalogram (EEG) although hurting human skin in long time monitoring. Moreover, wet electrode made of stainless steel materials with good corrosion resistance is better than dry electrode in a short time. Stainless steel materials would be corroded by sweat and cause electrochemical noise which can be solved by plating a layer of film [18] .

4. FEEDBACK PHYSICAL CONDITION BY CLASSIC ECG

Normal ECG performance reflects normal human physiological characteristics. As shown in Figure 2, biolo- gical signals collected by fabric electrodes will be compared with classic ECG after filtering and amplifying processing. Any erratic performance appeared in psychical condition would appear in one band of classic signals. This performance provides a good basis for arrhythmia, cardiac tumors, heart disease and so on.

To improve the fabric electrodes can meet the requirements of medical electrodes. Sun Jiaxing [19] et al. acquired ECG signals by carbon nano tube fabric electrodes and compared them with standard ECG signals. In a

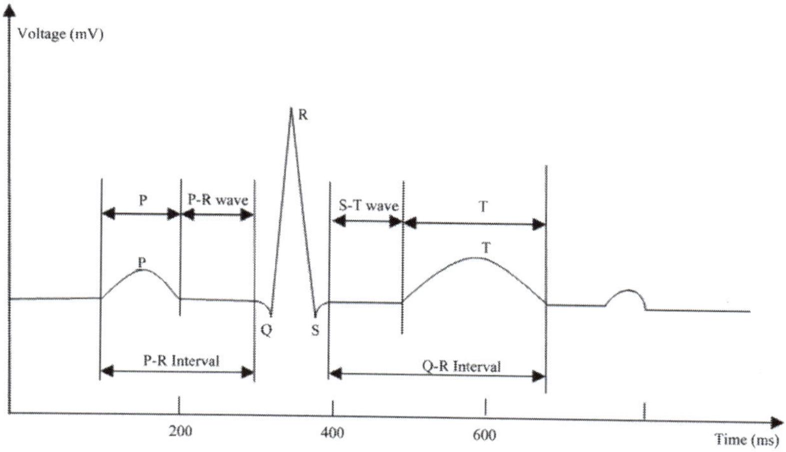

Figure 2. Classic electrocardiogram.

Table 2. Parameters of electrocardiogram.

Type of wave	Cycle	Amplitude	Source
P-wave	0.06 - 0.11	0.05 - 0.25	Produce left and right atrium signal
Q-wave	<0.03 - 0.04	<R 1/2 - 1/4	Deliver left and right atrium signal
R-wave	-	<2.5	Deliver left and right atrium signal
S-wave	0.06 - 0.11	-	Deliver left and right atrium signal
T-wave	0.05 - 0.25	0.1 - 1.5	Electric potential of ventricular repolarization
P-wave	0.06 - 0.14	Coplanar with the baseline	Between P-wave and QRS-weave
PR interval	0.12 - 0.20	-	Deliver to ventricular
ST-wave	0.05 - 0.15	Line	Process of ventricular restoration
QT interval	<0.4	-	process of ventricular repolarization

stationary state, PR interval, QRS interval and QT interval change little, the amplitude of Q-wave and S-wave and the error rate of characteristic waves are larger. Under the motion of standing, walking and jogging, PR interval is reduced and the error rate of characteristic waves is much larger. But the QRS interval and QT interval change little and the quality of signals remains clear.

Resting potential and action potential in medicine are known as polarization and depolarization. A classic electrocardiographic includes not only P-waves produced by atria depolarization, QRS-waves produced by ven- tricular depolarization, T-waves produced by rapid ventricular repolarization and so on. At the moment of atria depolarization, the beating heart could produce biological signals of P-waves. QRS-waves can also be produced by ventricular depolarization. T-waves, PR segment, ST segment, PR interval and ST interval can appear at the time of rapid ventricular repolarization. Classic ECG can accurately reflect the changes of biological signals and provide a reference for medical treatment [20] - [23] . Every band corresponds to a kind of mechanical moment of ventricular or atria as shown in Table 2.

5. CONCLUSION

The development of fabric electrodes not only solved the problem that the skin was damaged by inserting tra- ditional needle electrode, but also solved the problem of skin irritation caused by conductive gel on surface electrode. Different from traditional electrode, fabric electrode has many advantages of inner and outer. The characteristics of comfortable, breathable, and non-irritating properties of textile materials are greatly favored by many researchers. It is in accord with the requirements of medical and health field oriented toward miniaturization, intelligence and convenience, and fabric electrode has become one of the hottest researches among researchers in this area. As every coin has two sides, fabric electrode is not perfect because the biological signals acquired by fabric electrode need future processing. The selections of structure and materials are the most important factors. For the biological signals collected, how to filter the noise, how to reduce sliding artifacts and other problems are the other hot issues needed to be resolved for fabric electrodes. In a word, fabric electrode as a kind of dry electrode could meet the modernization medical needs. There will be a great market demand of fabric electrode. Therefore, fabric electrode has a high research value and it will also bring social and economic benefits.

ACKNOWLEDGEMENTS

Financial supports from Shanghai Municipal Education Commission Project (15 cxy36) and Innovative Project 2014-zx-01-2 of Shanghai Textile Co., LTD. are gratefully acknowledged.

REFERENCES

1. Di Rienzo, M., Rizzo, F., Parati, G., Brambilla, G., Ferratini, M. and Castiglioni, P. (2005) MagIC System: A New Textile-Based Wearable Device for Biological Signial Monitoring Applicability in Daily Life and Clinical Setting. 27th Annual International Conference of the Engineering in Medicine and Biology Society, Shanghai, 7167-7169.

2. Paradiso, R., Loriga, G., Taccini, N., Gemignani, A. and Ghelarducci, B. (2005) WEALTHY—A Wearable Healthcare System: New Frontier on e-Textile. Journal of Telecommunications and Information Technology, 4, 105-113.

3. Habetha, J. (2006) The MyHeart Project-Fighting Cardiovascular Disease by Prevention and Early Diagnosis. 28th Annual International Conference of the IEEE EMBS Conference, New York, 30 August-3 September 2006, 6746-6749.

4. A Wearable Wireless Vital Signs Monitor for Patients' Mobility in Cardiology and Sports. 4th International Conference on-NO PERMISSIONS Pervasive Computing Technologies for Healthcare (Pervasive Health).

5. Rosso, R., Munaro, G., Salvetti, O., Colantonio, S. and Ciancitto, F. (2010) CHRONIOUS: An Open, Ubiquitous and Adaptive Chronic Disease Management Platform for Chronic Obstructive Pulmonary Disease (COPD), Chronic Kidney Disease (CKD) and Renal Insufficiency. 32nd Annual International Conference of the IEEE EMBS.

6. Pereira, T., Carvalho, H., et al. (2013) Wearable Biopotential Measurement Using theTIADS1198 Analog Front-End and Textile Electrodes. IEEE Signal Conditioning and Signal Quality Assessment, 325-330.

7. Lourenco, A., Silva, H., Carreiras, C. and Fred, A.L.N. (2013) Outlier Detection in Non-Intrusive ECG Biometric System. Proceedings of the International Conference on Image Analysis and Recognition (ICIAR), 43-52.

8. Zhang, H., Tao, X., Tao, W., Li, W., Xu, P. and Liu, H. (2011) Textile-Structured Human Body Surface Biopotential Signal Acquisition Electrode. 4th International Congress on Image and Signal Processing (CISP), Shanghai, 15-17 October 2011, 2792-2797.

9. Zhang, Y. (2014) Study on Wearable Ambulatory Elecotrocardiogram Monitoring System and Electrocardiosignal Processing Method. University of Electronic Science and Technology of China.

10. Salvado, R., Loss, C., Gonçalves, R. and Pinho, P. (2012) Textile Materials for the Design of Wearable Antennas: A Survey. Sensors, 12, 15841-15857.

11. Liang, G., Zhu, L., Xu, J., Fang, D., Bai, Z., Xu, W. (2013) Investigations of Poly(pyrrole)-Coated Cotton Fabrics Prepared in Blends of Anionic and Cationic Surfactants as Flexible Electrode. Electrochimica Acta, 103, 9-14.

12. Zhai, H., Wang, C., et al. (2012) ECG Signal Monitoring System Based on Textile Electrodes. Journal of Jilin University (Information Science Edition), 30, 185-191.

13. Tao, D. (2011) Study of Textile Electrodes and Signal Processing Technology in Myoelectric Prostheses Control. Harbin Institute of Technology, Harbin.

14. Zhang, Y. (2014) Study on Wearable Ambulatory Elecotrocardiogram Monitoring System and Electrocardiosignal Processing Method. University of Electronic Science and Technology of China, Chengdu.

15. Yue, B., Wang, C., Ding, X. and Wallace, G.G. (2013) Electrochemically Synthesized Stretchable Polypyrrole/Fabric Electrodes for Supercapacitor. Electrochimica Acta, 113, 17-22.

16. Fu, X., Zhao, Y. and Cai, Z. (2010) Preparation of Polyaniline/Metal Composite Conductive Fabric by Electroless Plating. Dieing Polymerica Sinia, 36, 1-4.

17. Jing, X., Wang, L., et al. (2005) Synthesis, Structure, Properties and Applications of Conducting Polyaniline. Acta Polymerica Sinia, 655-663.

18. Webster, J. (2009) Medical Instrumentation: Application and Design. Wiley, India.

19. Sun, J., Wu, Y., et al. (2010) Wearable Dynamic ECG Monitoring System Based on the Signal Quality Evaluation. Donghua University, Shanghai.

20. Zhang, Z. (2010) Research on Acquisition of Electromyography for Prosthesis Control. Harbin Institute of Technology, Harbin.

21. Xi, X., Zuo, J., et al. (2014) A Study of Multi-Channel EMG De-Noising and Aliasing Removal. Chinese Journal of Sensors and Actuators, 293-298.

22. Liang, G., Zhu, L., Xu, J., et al. (2013) Investigations of Poly(pyrrole)-Coated Cotton Fabrics Prepared in Blends of Anionic and Cationic Surfactants as Flexible Electrode. Electrochimica Acta, 103, 9-14.

23. Zhang, L., Yan, H., et al. (2014) Application and Research Progress in ECG Signal Detection of Non-Contact Electrode. Space Medicine & Medical Engineering.

11

CHAPTER

Statistical Analyses and Predicting the Properties of Cotton/Waste Blended Open-End Rotor Yarn Using Taguchi OA Design

Md. Khalilur Rahman Khan[1], Mohammad Mobarak Hossain[1], Ronobir Chandra Sarker[2]

[1]*Department of Textile Engineering, Bangladesh University of Business and Technology (BUBT), Dhaka, Bangladesh*

[2]*Department of Mathematics, Bangladesh University of Business and Technology (BUBT), Dhaka, Bangladesh*

ABSTRACT

Quality and cost of production are two important considering parts during production of cotton/waste blended open-end rotor spun yarn. Nowadays, blending of spinning waste with virgin cotton is a common phenomenon in rotor spinning intended to lower the production cost. As a result, understanding the relationships between properties of cotton/waste blended rotor yarn and different spinning conditions are important for spinner to maintain acceptable quality. However, this paper focuses on predicting the properties of cotton/waste blended rotor yarn, using taguchi OA experimental design. In this study, the experimental quality characteristics of 16 Ne rotor yarn focused on are the, unevenness, imperfections, strength and elongation wherein before blend ratios, blending technique, two process parameters namely cylinder speed and rotor speed were used as predictors. Using ANOVA, the significant factors influencing the quality characteristics of rotor spun yarn are obtained and analyzed by interaction matrix of dependent variables. Result shows that the most influential parameter in our investigation was waste proportion as expected. Though inferior in strength, the benefits of draw frame blending in OE spinning may be more than blow room blending in terms of reducing the imperfection, irregularity and yarn elongation during processing of higher waste mixing. With increasing rotor speed, yarn evenness, imperfection and strength show a significant deterioration, in case of elongation, no consistent trend was observed.

Keywords: Rotor speed, Cylinder speed, Blending technique, Rotor yarn, Waste, Taguchi design

1. INTRODUCTION

Many open-end spinning methods have been invented, but none have been successful than rotor spinning [1]. The volume of production on rotor spinning has increased in recent years which is quite understandable considering the present trend in the production and consumption of textile products [2].

Improving raw material exploitation has become the most important challenge facing scientific and industrial community [3]. Extreme competition in textile world have forced the textile mills to reduce the production cost that is directly connected with raw material cost. In spite of the technical evolution for different blow room machines, generated wastes in cotton spinning mill contain a great rate of fibers. [4] Waste utilization in textile industry of our country has gained vital importance because of high price in cotton market. Some researcher has been worked on reuse of recycled cotton waste [5-7].

Various researchers have studied the effect of rotor speed and carding parameters on rotor yarn quality [8-12]. It is also observed that many researchers have studied rotor spinning process and yarn quality from different angle. No work has so far been observed on the influences of process parameters viz. cylinder speed and rotor speed for different blending technique on rotor blended yarn quality produced by using high amount of recycled wastes.

Blending is used as a means of substituting a less expensive fiber for a highly priced one and to produce yarns with qualities that cannot be obtained by using one type of material alone [13]. Blending virgin cotton fibers with recycled wastes is a widely practiced means of enhancing the performances and the desired qualities of rotor yarn in spinning mills. Different blend ratios for blended rotor spun yarn produced form virgin cotton and recycled spinning wastes are used in textile mill.

Recovered secondary raw materials can be blended with virgin cotton fibers either at blow room or at draw frame. Generally, it is said that the earlier blending of fibers give better blending. Most of the spinning mill in our country practice hand mixing of virgin cotton and recycled wastes in blow room. As a result, with huge spaces, a large number of workers are needed that add extra cost on production. On the other hand, sliver of 100% virgin cotton and sliver of 100% recycled wastes can be blended easily at finisher draw frame without extra space and labor cost to produce rotor yarn with different level of blend proportion.

However, the textile industry is one of the largest and oldest industries worldwide and yarn manufacturing is the key process of it [14]. In general, the various responses in yarn manufacturing depend on a number of independent variables. So, collecting data from improperly designed experiments and analyzing them cannot help the researcher to make appropriate inferences.

The complexity of fiber to yarn spinning process and interaction among various spinning conditions is very high. Design of Experiment (DOE) is an experimental or analytical method that is commonly used to statistically signify the relationship between input parameters to output responses, where by a systematic way of planning of experiments, collection and analysis of data is executed [15]. Statistical models were the first that used in textile disciplines to explore relationships between variables and characteristics of product and to optimize processing parameters [16]. Using experimental design as a prediction approach with higher degree of precision is a subject focused by many researchers. The experimental design proposed by Taguchi involves using orthogonal arrays to organize the parameters affecting the process and the levels at which they should be varies. Instead of having to test all possible combinations like the factorial design, the Taguchi method tests pairs of combinations. This allows for the collection of the necessary data to determine which factors most affect product quality with a minimum amount of experimentation, thus saving time and resources [17].

Economical running behavior can only be achieved by keeping in observation the effects of process parameters on yarn properties and raw material [18]. However, it is a critical problem to optimize the spinning parameters for example cylinder speed, rotor speed etc. to produce rotor yarn using high amount of recycled wastes for both blow room blending and draw frame blending.

The work reported here is concerned about the investigation of the rotor yarn quality responses in terms of yarn regularity, imperfections, and strength and elongation by using Taguchi OA Factorial design of experiment. To conduct the tests using the taguchi approach two levels for each dependent variable (Table-1) were selected.

Table 1. Selected independent variables

Variable	Value/Type	
Blend Ratio	67/33	83/17
Blending Technique	Blow Room	Draw frame
Cylinder Speed (rpm)	800	850
Rotor Speed	80,000	85,000

2. MATERIALS AND METHODS

2.1. Preparation of Yarn Samples

In this study, Flat strips (F/S) collected from carding machines during processing in spinning mill are recycled by recycling machine before pressing by bale press machine. Comber noils were collected from comber machine and pressed to form bale of comber noil. Unused packages of roving were processed at Rovematric frame to collect untwisted roving wastes and Pneumafil wastes were collected from ring frame directly without any further processing.

However, Shankar-6 fiber is used as virgin cotton to blend with spinning waste. Properties of Waste materials and Shankar-6 have been given in Table-3. Rotor yarns of 16 Ne were spun from virgin cotton and recycled wastes with two different levels of blend proportion, viz. 17:83 and 33:67 for both blow room blending and draw frame blending. In case of blow room blending, hand-mixing concept was applied in blow room to blend virgin cotton and recycled wastes. On the other hand, during draw frame blending, carded sliver produced from recycled wastes was blended with the carded sliver produced from 100% virgin cotton in finisher draw frame (RSBD-22) and blend ratio have been maintained by producing sliver of different thickness. Compositions of waste material and processing conditions of rotor spun yarn have been given in Table-2 and Table-4.

Table 2. Composition of waste materials used in blow room mixing

Blend ratio at percentage (Virgin cotton/Spinning waste)	Waste type	Percentage of total waste	Percentage in mixing
17/83	Flat strips (F/S)	65%	54
	Comber noil	15%	12.5
	Pneumafil	10%	8.3
	Roving	5%	4.1
	Filter waste	10%	8.3
33/67	Flat strips (F/S)	65%	43.6
	Comber noil	20%	13.4
	Pneumafil	10%	6.7
	Roving	5%	3.3

Table 3. Properties of spinning waste used in mixing

Fiber properties	Shankar-6	Spinning waste
Average Fineness (Mic value)	4.1	3.8
Average length (mm)	26.8	22.5
Length CV%	19.2	32.9
Fiber tenacity (cN/tex)	29.4	25.6
Maturity ratio	0.90	0.87
Trash Area (%)	0.15	0.28

Table 4. Processing conditions of rotor spun yarm

Machine	Parameter	Value
UNIclean , B12	Cleaning Intensity (C.I)	0.4
	Relative amount of waste (R.W)	4
UNIflex, B60	Cleaning Intensity (C.I)	0.3
	Relative amount of waste (R.W)	3
Carding, C60	Taker-in speed	1200 rpm
	Flat speed	0.30 m/min
	Cylinder speed	800 & 850 rpm
	Licker in to cylinder setting	0.22mm
	Cylinder to flat settings	0.25 mm
	Cylinder to doffer setting	0.2 mm
	Delivery rate	70 kgs/hr
	Carded sliver hank	91 gr/yd
Draw Frame, RSB D-22	Total draft	8.2
	Break draft	1.15
	Delivery speed	450 m/min
	Drawn sliver hank	75 gr/yd
Rotor frame, BT 923	Rotor speed	80000 & 85000 rpm
	Rotor diameter	33mm
	Opening roller speed	9000 rpm
	Navel	Ceramic & smooth

For both blending techniques, Rieter C-60 card machine and spinning frame BT-923 have been used. Carding production rate kept equal (70kgs/hr) for both blending techniques. Rotor speed was selected 80000 rpm and 85000 rpm, cylinder speed was selected 800 rpm and 850 rpm for their higher value and lower value used in our experimental design. Taguchi OA Factorial experimental design has been used in the investigation. 10 yarn samples were produced for each experimental set up according to Taguchi OA Factorial design. ANOVA table and main effect plots have been used to identify the significant processing conditions. The cut-off value 0.1 for the p value has been used to evaluate significant factors meaning when p value less than 0.1, then the null hypothesis is rejected. As a result, the factors will have influencing effect on the responses.

2.2. Testing of Yarn Samples

The USTER tester-5 was used to measure the evenness of yarn and imperfection/km. Yarn imperfection has been calculated by adding Thick place(+50%), Thin place(-50%) and Neps (+280%) per km. Yarn testing speed was 400m/min. Average results of 10 samples produced for each experiment according to design have been calculated. Single yarn strength and yarn elongation% have been measured by using MAG Elestretch single yarn strength tester.

3. RESULTS AND DISCUSSIONS

The experimental design and statistical analyses were performed using the Reliasoft. Quality parameters of 16 Ne rotor yarn produced at different level of independent variables are given in Table-5. Analysis of variance tables for selected responses viz. irregularity, imperfection, strength and yarn elongation are given in Tables 7, 8, 9 & 10 respectively. The regression co-efficient and p-values of all the terms are given in table-6. The impact of all the variables on the responses is discussed separately in the following sections:

Yarn Irregularity: From AONVA table of yarn irregularity, it can be said that our selected parameters have evidently two way interactions on yarn irregularity. Table-3 reveals that blend ratio (A), blending technique (B), cylinder speed (C) and the term A: B are statistically significant terms. From interaction matrix shown in figure-1, blend ratio 67/33 (virgin/recycled) produce always better result of yarn irregularity as we expected. Higher percentage of short fibers will hinder the fiber orientation in sliver during drafting process because of behaving as swimming or uncontrolled fiber.

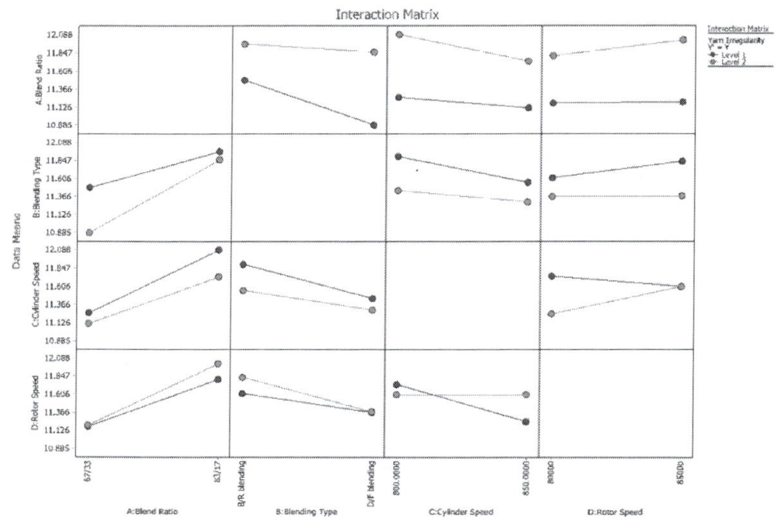

Figure 1. Interaction matrix for predicting yarn irregularity

Table 5. Experimental set up for 16 experiments according to Taguchi OA Factorial design

Expt No.	Blend Ratio	Blending Type	Cylinder Speed	Rotor Speed	Irregularity, U%	Imperfections, IPI/km	Strength, g/tex	Elongation, %
1.	83/17	D/F	850	85000	11.92	394	9.1	5.88
2.	67/33	D/F	800	85000	10.85	319	9.49	4.9
3.	83/17	B/R	800	85000	12.19	510	9.18	5.92
4.	83/17	D/F	850	85000	11.65	440.5	8.89	6.24
5.	67/33	D/F	850	80000	10.62	197.5	9.76	5.58
6.	67/33	B/R	800	80000	11.47	239.5	10.01	5.71
7.	83/17	B/R	800	85000	12.3	461	9.25	5.65
8.	67/33	D/F	800	85000	11.08	268.4	9.64	5.03
9.	83/17	B/R	850	80000	11.63	422.5	9.34	5.22
10.	83/17	D/F	800	80000	11.77	411.1	9.16	6.13
11.	67/33	B/R	800	80000	11.63	282.4	9.84	6.17
12.	67/33	B/R	850	85000	11.5	301	9.67	5.07
13.	83/17	B/R	850	80000	11.74	439.2	9.26	5.44
14.	83/17	D/F	800	80000	12.09	450.3	8.76	6.19
15.	67/33	D/F	850	80000	10.99	297.2	9.65	5.21
16.	67/33	B/R	850	85000	11.35	331.8	9.52	4.83

Table 6. Regression Information for dependent variables

Term	Coefficient				P Value			
	Irregularity	Imperfections	Strength	Elongation	Irregularity	Imperfection	Strength	Elongation
Intercept	11.5488	360.3375	9.4075	5.5731	0	1.98E-10	0	5.88E-14
A: Blend Ratio	-0.3625	-80.7375	0.29	-0.2606	2.24E-05	2.23E-05	2.93E-05	0.001
B: Blend Type	0.1775	13.0875	0.1012	-0.0719	0.0026	0.1927	0.0184	0.1999
C: Cylinder Speed	-0.1238	-7.375	-0.0088	-0.1394	0.0172	0.4459	0.8051	0.0267
D: Rotor Speed	-0.0563	-17.875	0.065	0.1331	0.2107	0.0879	0.0948	0.0322
A:B	0.1237	-4.0125	-0.0388	0.2044	0.0172	0.6743	0.2915	0.0041
A:C	0.0525	9.65	-0.0387	-0.0006	0.2397	0.3249	0.2915	0.9906
A:D	0.0475	-7.575	0.0525	0.2219	0.2837	0.4342	0.1645	0.0026

Table 7. ANOVA Table for Yarn Irregularity

Source of Variation	df	Sum of Squares (Partial)	Mean of Squares	F Ratio	P Value
Model	7	3.22	0.461	16.86	0.0003
Main Effects	4	2.90	0.725	26.54	0.0001
2 Way Interaction	3	0.32	0.108	3.965	0.0529
Residual	8	0.21	0.027		
Pure Error	8	0.21	0.027		
Total	15	3.44			

Table 8. ANOVA Table for Yarn Imperfection

Source of Variation	df	Sum of Squares (Partial)	Mean of Squares	F Ratio	P Value
Model	7	1.16E+05	1.65E+04	12.20	0.0011
Main Effects	4	1.13E+05	2.83E+04	20.86	0.0003
2 Way Interaction	3	2665.652	888.5508	0.65	0.6015
Residual	8	1.08E+04	1354.2675		
Pure Error	8	1.08E+04	1354.2675		
Total	15	1.27E+05			

Table 9. ANOVA Table for Yarn Strength

Source of Variation	df	Sum of Squares (Partial)	Mean of Squares	F Ratio	P Value
Model	7	1.670	0.2387	12.6	0.0009
Main Effects	4	1.578	0.3946	20.9	0.0003
2 Way Interaction	3	0.092	0.0307	1.63	0.2577
Residual	8	0.150	0.0188		
Pure Error	8	0.150	0.0188		
Total	15	1.821			

Table 10. ANOVA Table for Yarn Elongation

Source of Variation	df	Sum of Squares (Partial)	Mean of Squares	F Ratio	P Value
Model	7	3.2198	0.46	10.86	0.001
Main Effects	4	1.7638	0.441	10.41	0.002
2 Way Interaction	3	1.456	0.485	11.46	0.002
Residual	8	0.3388	0.042		
Pure Error	8	0.3388	0.042		
Total	15	3.5585			

In comparison of blending techniques, draw frame blending gives lower irregularity than blow room blending technique. This statement is true prominently for blend ratio 67/33.It may be attributed because of separate processing of virgin cotton and recycled wastes in card during draw frame blending technique. The task of individualization and parallelization of virgin cotton performed by card cylinder hampers due to excessive recycled wastes in case of blow room blending. On the other hand, blending technique hardly improves the evenness of rotor yarn at blend ratio 83/17. Results show that higher cylinder speed and smaller rotor speed give better results in term of yarn evenness. The trend in the results can be explained as follows: higher cylinder speed creates higher carding force per unit load. Yarn irregularity (U %) decreases with the increase of carding force [19]. On the other hand, feed rate have to increase to maintain the yarn count for a given yarn twist value with the increase of rotor speed. As a result, the efficiency of combing roller in individualizing the fibers gets decreased. Ultimately yarn evenness will be less in high rotor speed in most cases. The trend is in agreement with Halimi et al. [20]. Same trend of results are found also by F.A Arain et al. and they explained that too high rotor speed results in too much increase in number of wrapper fibers and disturbs the fiber arrangement in the groove of rotor, so as a result unevenness of yarn increases [21].

Yarn Imperfections: Our selected experimental design model is statistically significant that have a p-value of $0.0011<0.1$. From regression information given in table-3, we get blend ratio as a significant term for yarn imperfection. From interaction matrix, the influences of different factors can be stated as follows: high content of virgin cotton in blend composition leads to lesser yarn imperfections value and also at the same blend ratio, blending technique or carding speed have very little influence on yarn imperfections. It is well established that high amount of wastes used in processing deteriorates yarn quality such as yarn imperfection value. Among three independent variables viz. blending technique, cylinder speed and rotor speed, yarn imperfection results be influenced by rotor speed ($p<0.1$) more than other two parameters considered individually. With the increase of rotor speed, yarn imperfection value also increase. These results happened because of higher rotor speed forces the wrapper fibers to be closely wrapped around the yarn body. This type of wrapper fibers is counted as yarn neps. Fibers generally contain hooks or loops, have less time to be aligned themselves in rotor groove before yarn formation at high rotor speed. As a result, yarn imperfection value increase with the increase of rotor speed.

Yarn Strength: Yarn strength, is one of the most important parameter influencing the yarn's use in terms of quality. ANOVA table for yarn strength has been given in table-4. It is clear from ANOVA results of yarn strength that p value of design model and main effects indicating the significance of selected experimental design. However, according to table-3 results the independent variables blend ratio, blending technique and rotor speed have a significant effect ($p<0.1$) on the strength of yarn. Whereas cylinder speed has not the same effect ($p>0.1$). Higher percentage of virgin cotton always increase yarn strength. From interaction matrix of yarn strength (figure-3), the value of strength has been decreased slightly with the increase of rotor speed almost for any levels of other independent variables. Fibers coming out of combing roller cannot be straightened in rotor groove because of getting short time before yarn formation which results in lower yarn strength. From interaction matrix it is also seen that during blow room blending, yarn strength has been increased for any level of other variables. Homogeneity in blending is necessary to get higher yarn strength during blending of recycled waste and virgin cotton. Blending in blow room was found to be better, resulting in more homogeneous mixing of the component fibers [22]. Strength increased with high cylinder speed in case of draw frame blending whereas opposite results fount in case of blow room blending. For blow room blending, neps reduction and the effect of individualization will be less because of processing recycled waste and virgin cotton together. Virgin cotton blended with recycled wastes will face unnecessary stresses applied during processing at blow room since more cleaning actions are needed in case of recycled wastes. As a result processing virgin and recycled cotton separately in carding machine can avoid disturbances.

Yarn Elongation: According to p-value the independent parameters can be classified according to the level of important as: A (blend ratio)> A: D>A: B> C (cylinder speed)>D (rotor speed). Figure-4 shows the interaction effects of different factors on yarn elongation. Results indicate that yarn elongation drops at draw frame blending and higher rotor speed markedly for blend ratio 67/33 (recycled/virgin) but increases for blend ratio 83/17. It may be due to the presence of more short fibers in blend composition of 83/17 create disturbance during drafting. Parallelization of fiber in sliver will be poor that will be resulted in higher elongation of rotor yarn because of having hooked, curled fiber. On the other hand, high amount of centrifugal force will be acted on fiber during high rotor speed. As a result, yarn structure will be compact and fiber slippage will be decreased. Ultimately yarn elongation will loss at high rotor speed.

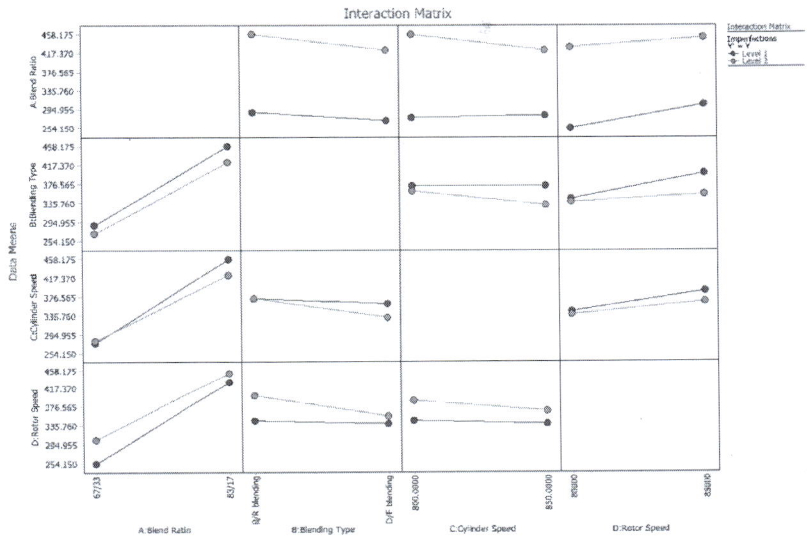

Figure 2. Interaction matrix for predicting yarn imperfection

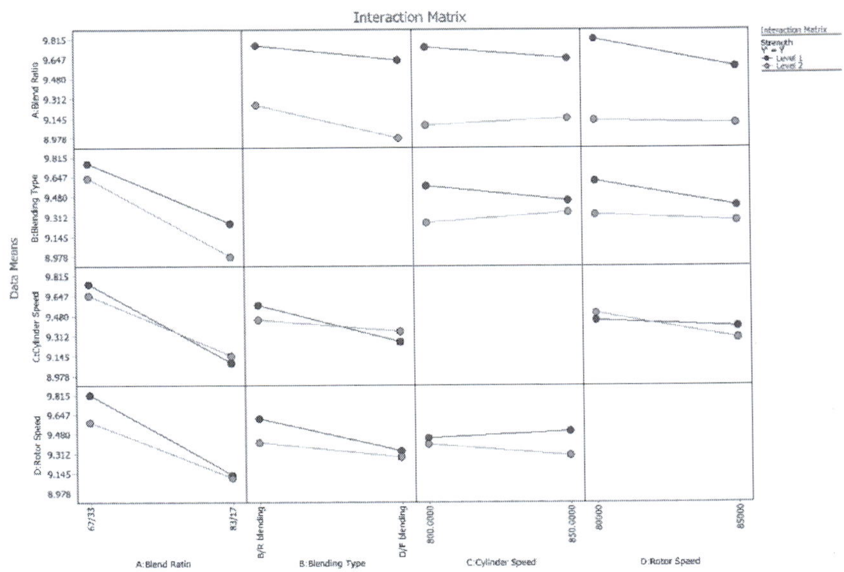

Figure 3. Interaction matrix for predicting yarn strength

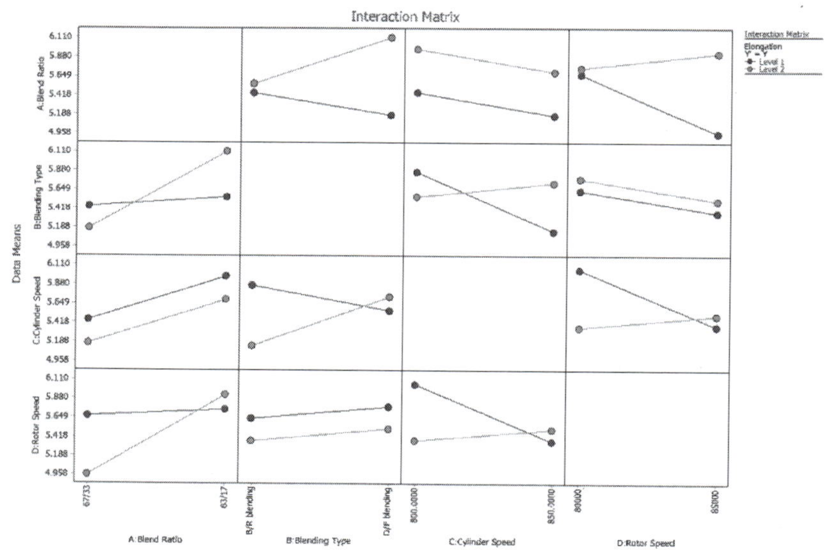

Figure 4. Interaction matrix for predicting yarn elongation

Cylinder speed seems to have a significant effect on yarn elongation. Working with high cylinder speed affects negatively with yarn elongation. Fibers get better orientation and straightening due to high carding force at high cylinder speed resulted in better drafting operation at finisher draw frame. So, fiber separation by opening roller in spinning frame will be easier. Better separation by opening roller will lead lower elongation in rotor yarn.

4. CONCLUSIONS

In this study we have tried to predict the most important quality variables of rotor spun yarn based on four independent variables by using taguchi OA factorial design. For this purpose, 16 samples of 16 Ne rotor yarn have been produced. From ANOVA results, it is clear that the significance level of regression (p value) for our selected model is less than 0.01 meaning model is significant to predict the quality of rotor spun yarn. Our model pointed out that blend ratio and rotor speed are most influencing factors for producing good quality yarn. Reducing rotor speed is beneficial to the yarn evenness, imperfections and strength. Although the shortcoming of draw frame blending is lower yarn strength but we can improve yarn quality produced from higher mixing of recycled waste in terms of evenness, imperfection and elongation by using draw frame blending in place of blow room blending. A high cylinder speed is recommended at draw frame blending and during blending of higher waste, while spinning waste content does not provide the expected results of yarn strength.

REFERENCES

1. Rameshkumar C.1, Anandkumar P.1, Senthilnathan P.1, Jeevitha R.1, Anbumani N.2, "COMPARITIVE STUDIES ON RING ROTOR AND VORTEX YARN KNITTED FABRICS", AUTEX Research Journal, December 2008, Vol. 8, No4, © AUTEX,pp-100.

2. S M Ishtiaque, "Spinning of synthetic fibres and blends on rotor spinning machine", Indian Journal of Fibre & Textile Reasearch, December-1992,Vol-17, pp-224-230.

3. Halimi Mohamed Taher, Ph.D., Azzouz Bechir, Ph.D., "Ben Hassen Mohamed, Sakli Faouzi, Influence of Spinning Parameters and Recovered Fibers from Cotton Waste on the Uniformity and Hairiness of Rotor Spun Yarn",Journal of Engineered Fibers and Fabrics, Volume 4, Issue 3 – 2009, pp-36-44].

4. Hanafy, "Fabric from cotton waste", The Indian Textile Journal, pp. 16–18, 1997.

5. M. Taher Halimia, M. Ben Hassena, B. Azzouza & F. Saklia, "Effect of cotton waste and spinning parameters on rotor yarn quality", Journal of the Textile Institute, Volume 98, Issue 5, 2007, pages 437-442 .

6. B. Wulfhorst, "The technological and economic aspects of the recycling of wastes in modern cotton mills", Textile Praxis International, p741-743, 1984, August.

7. Pinar N. Duru and Osman Babaarslan, "Determining an Optimum Opening Roller Speed for Spinning Polyester/ Waste Blend Rotor Yarns", Textile Research Journal, October 2003; vol. 73, 10: pp. 907-911.

8. J.S. Manohar, A.K. Rakshit, and N. Balasubramanian," Influence of Rotor Speed, Rotor Diameter, and Carding Conditions on Yarn Quality in Open-End Spinning", Textile Research Journal, August 1983; vol. 53, 8: pp. 497-503.

9. Jacic Simpson and Myles A. Patureau, "Effect of Rotor Speed on Open-End Spinning and Yarn Properties", Textile Research Journal, August 1979; vol. 49, 8: pp. 468-473.

10. K.R. Salhotra and P. Balasubramanian, "Influence of Carding Technique on Yarn Tenacity Response to Increasing Rotor Speed", Textile Research Journal, June 1985; vol. 55, 6: pp. 381-382.

11. Sengupta A K, Vijayaraghavan N & Singh A., "Studies on carding force between cylinder and flats in a card: part III-carding parameters, sliver quality and carding force", Indian J Text Res,8(1983)68.

12. Halimi Mohamed Taher, Ph.D., Azzouz Bechir, Ph.D., Ben Hassen Mohamed, Sakli Faouzi , "Influence of Spinning Parameters and Recovered Fibers from Cotton Waste on the Uniformity and Hairiness of Rotor Spun Yarn" ,Journal of Engineered Fibers and Fabrics, Volume 4, Issue 3 – 2009.

13. Babak Yadollah Roudbari and Safdar Eskandarnejad, "Effect of Some Navels on Properties of Cotton/Nylon66 Blend (1 : 1) Rotor Spun Yarn and Wrapper Formation: A Comparison between Rotor and Ring Spun Yarn", Journal of Textiles Volume 2013 (2013), Article ID 262635, Hindawi Publishing Corporation.

14. Lokesh Shukla, Anita Nishkam, "Performance Optimization, Prediction, and Adequacy by Response Surfaces Methodology

with Allusion to DRF Technique", ISRN Textiles, Hondawi Publication Corporations, Volume 2014 (2014), Article ID 634041.

15. S. P. Kondapalli, S. R. Chalamalasetti and N. R. Damera "Application of Taguchi based Design of Experiments to Fusion Arc Weld Processes: A Review", International Journal of Technology and Management. Vol. 2 No. 1, pp. 1-8 (2013).

16. Erol, R.; Sagbas, A.; Multiple Response Optimisation of the Staple-Yarn Production Process for Hairiness, Strength and Cost; Fibres and Textiles in Eastern Europe 2009, 17, 40-42

17. Fraley, Stephanie, et al. "Design of experiments via Taguchi methods: orthogonal arrays." The Michigan chemical process dynamic and controls open text book, USA 2.3 (2006): 4.

18. F. Ahmed , S. Saleemi , A. W. Rajput , I. A. Shaikh ,A. R. Sahito ,"Characterization of Rotor Spun Knitting Yarn at High Rotor Speeds", Technical Journal, University of Engineering and Technology (UET) ,Taxila, Pakistan Vol. 19 No. IV-2014.

19. F. Ahmed , S. Saleemi , A. W. Rajput , I. A. Shaikh ,A. R. Sahito ,"Characterization of Rotor Spun Knitting Yarn at High Rotor Speeds", Technical Journal, University of Engineering and Technology (UET) Taxila, Pakistan Vol. 19 No. IV-2014.

20. F.A Arain, A. Tanwari, & H.R. Sheikh ,"Statistical Modeling for the Effect of Rotor Speed, Yarn Twist and Linear Density on Production and Quality Characteristics of Rotor Spun Yarn", Mehran University Research Journal of Engineering & Technology, Volume 31, No. 1, January, 2012.

21. Farooq Ahmed Arain, Anwaruddin Tanwari, Tanveer Hussain and Zulfiqar Ali Malik, "Multiple Response Optimization of Rotor Yarn for Strength, Unevenness, Hairiness and Imperfections", Fibers and Polymers 2012, Vol.13, No.1, 118-122.

22. T.V. Ratnam and K.P. Chellamani, "Quality control in spinning, page-59, Published by The South India Textile Research Association, Coimbatore, 1999.

12
CHAPTER

An Overview of Modeling Yarn's 3D Geometric Configuration

Tianyong Zheng, Jing Wei, Zhengtao Shi, Tingting Li, Zhen Wu

School of Textile Engineering, Zhongyuan University of Technology, Zhengzhou, China

ABSTRACT

Modeling the 3D geometric configuration of yarn is the base of building the 3D fabric structure. The surfaces of the yarns are modeled by many facets. To improve the efficiency of modeling, 3D yarn is usually designed by Bezier curve, spline curve or B-spline curve on the principle of Computer Graphics. The paper reviews different 3D modeling methods for single yarn from the view of cross-section and central line. Some improvements for a better visual effect such as inserting twist and hairiness are illustrated as well. Based on the single yarn, the folded yarn and some fancy yarns can also be modeled in 3D space as long as the cross-sections of the control sections of the component single yarn are determined. Applications of the various 3D yarns are demonstrated. The advantages and the drawbacks of various methods are explained and compared as well. Comparatively speaking, modeling technology based on spline curve is more convenient and flexible, perhaps it is the most popular tool since it is capable of designing different kinds of yarns mentioned above and gets the maximum support from the different developing platforms such as OpenGL, 3DS-MAX, Solid Works CAD, etc.

Keywords: 3D Modeling, Yarn, Spline Curve, Computer Graphics

1. INTRODUCTION

The geometrical structures of the woven fabrics, knitted fabrics or braided fabrics affect the physical properties of the fabric to a great extent. These fabrics are composed of yarns, therefore, the 3D geometrical configuration of the yarn has to be modeled before the 3D geometrical structures of the fabrics are to be modeled. Various methods were proposed in modeling the 3D yarns and the paper aims to give a complete illustration of how to model a 3D yarn.

2. METHODOLOGY FOR SINGLE YARN

The configuration of 3D yarn is modeled according to Computer Graphics. Generally speaking, the surface of the 3D yarn can be regarded as composed of many tiny facets. As soon as all the facets of the yarn surface are determined, the normal of the each facet is determined accordingly, then the shading effect of the yarn will be calculated according to Phong or Goround [1] model, which gives the yarn the photorealistic effects. The key to build the 3D yarn is to design the facets or the vortexes of the facets or control polygons. Figure 1 demonstrates how a cylinder is built by the facets.

Obviously, it is a tedious work to construct an object by determining so many facets one by one. An efficient method must be found. Many surface modeling technologies were applied widely, such as Bezier surface, B- spline surface [2] as shown in Figure 2. For those surfaces, a grid determined by a serial of controlled vertexes in two directions is used to describe the surface. To accelerate the designing process, software such as OpenGL [3] , 3D MAX or DirectX which support B-spline surface or Bezier surface to facilitate the 3D model of an object and gives the object photorealistic effects as well.

To a 3D yarn model, the surface of the yarn can be regarded as a closed surface. Therefore, to construct a 3D model for yarn configuration, there are two issues that have to be considered: 1) representation of the central line; 2) description of the cross-section. Peirce [4] proposed a way to describe yarns in woven structure although his works were based on planar calculation. He thought the yarn is uniform in diameter and the cross-section of the yarn can be circular or elliptical. Kemp [5] and Hearle [6] proposed the cross-section of yarn can be of racetrack and lens depended on the structure. Hearle [7] also proposed that the central line of the yarn be cubic polynomial curve according to energy theory. The shape of the cross-section of the yare is suitable to knitted or braided fabric structure. It should be noted that there is still difference between the real yarn and those hypotheses, perhaps a new assumption will be introduced someday.

Based on these hypotheses, many 3D geometrical models of the yarn were setup. And these models will be introduced according to the method to control the shape of the central line or the cross-section. As for whether the cross-section is one of the shape stated above or how to calculate the true path of the central line by the fabric construction, which are beyond the research of the paper. What concerned in the paper are how to describe the shape of the central line of the yarn if some key points (or interpolated points) on the central line are determined and how to describe the special shape of the cross-section if the cross-section is given. Keep in mind that a good 3D model should not only be able to suit all those hypotheses available at present, but also to the ones may be proposed in future.

Figure 1.Constructing a cylinder by facets according to Computer Graphics [1] .

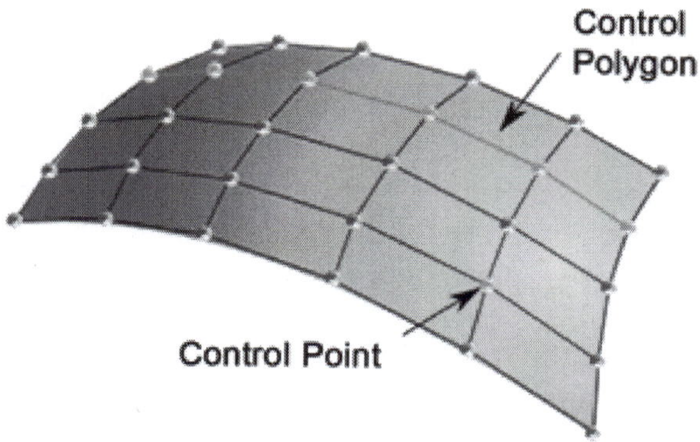

Figure 2. Control points for Bezier/B-spline surface: the shape of the control can be designed by shifting the control point [2] .

2.1. Model Based on Sweeping the Cross-Section along the Yarn Path

Liao and Adanur [8] put up a method to 3D modeling using a computer aided geometric design (CAGD). In this approach, the yarn is assumed to be a uniform cylinder which has a constant cross-section and the center line is an arbitrary spatial curve. The base cross-section is determined first, and it is divided into many arc by several points. A unique plane through any point on the center line and perpendicular to it can be defined. By sweeping the base cross-section curve along the points on the center line, all the points that determine the quadrilateral facets are automatically generated by coordinate transformation. The principle is well explained in Figure 3. Thus, the two 2D curves (yarn cross-section and yarn central line) are converted into a 3D surface that determines the shape of the object. In the model, the path of the yarn can be the combination of sinusoidal curve and segment line while the cross-section of the yarn can be circular, elliptical and race-track shapes which depend on the construction of the fabric. The structure of 3D woven fabric and braided fabric have been established. However, the cross-section of the yarn is uniform, which is unable to describe the true fabric structure and limits the application.

Lomov [9] proposed a 3D yarn model similar to Liao's model, and software Wise Tex is developed based on Lomov's work. Figure 4 demonstrates Lomov's research. Lin and Long [10] made use of the similar technique to model the yarn comprising of many individual sections of regular shape such as ellipse, lens that can be locally edited to avoid yarns penetrating when used in fabric structure.

2.2. Model of Uneven Straight Yarn

In order to reflect the effects of the uneven yarn, Li [11] builds a yarn which is made up with circles which has different diameters. The diameters can be acquired from the yarn uniformity tester. Segmentation on the surface of the yarn will be carried out by the evenly division of each circle along the center line of yarn. Capitalized on the number of the segment peaks, the contour of the yarn can be obtained by the list of the surface, edge and peak. This method is suitable to straight yarn and it can well reflect the variation of the cross-section. Figure 5 shows the 3D yarn contour in this method. However, the model can only show the straight yarn and cannot be applied to the yarn in fabric structure.

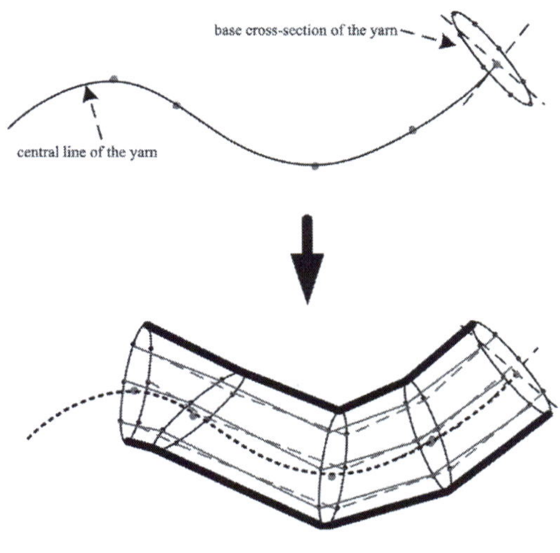

Figure 3. Principle of Liao's yarn model by sweeping the cross-sections along the yarn path [8] . Each cross-section of the yarn is perpendicular to the center line. All the points controlled the quadrilateral facets are automatically generated.

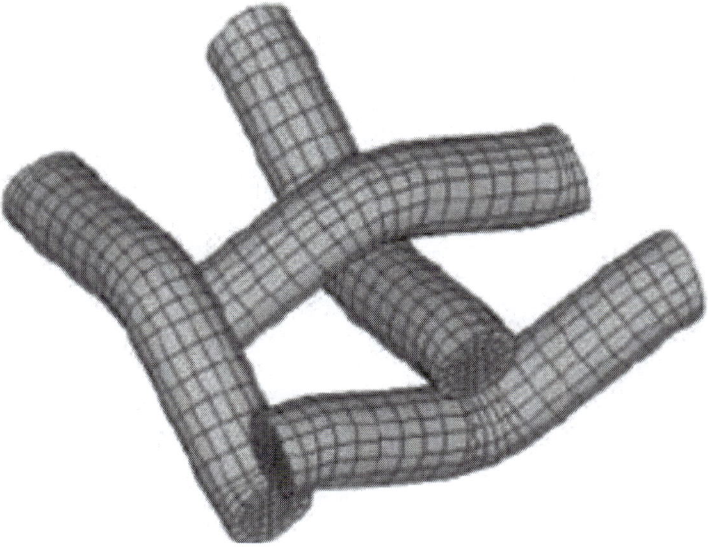

Figure 4. Lomov's work for 3D yarn [9] .

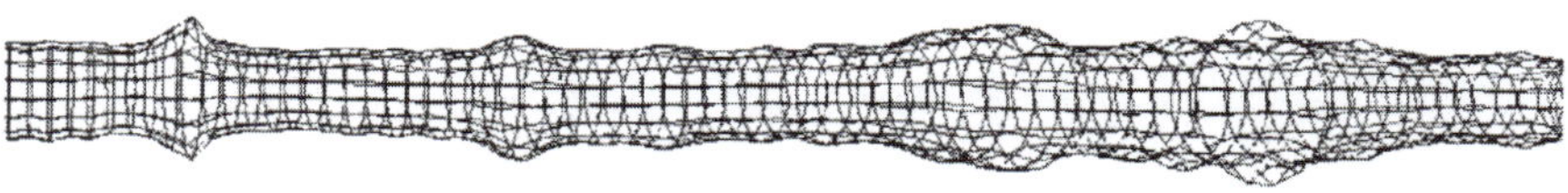

Figure 5. Li's yarn model with various cross-sections along the central line [10] .

2.3. Based on Polynomial Curve

Much work has been done to represent the different shapes of the yarn for the changing of shapes varies due to the flattening and strain. Gong and Ozgen [12] [13] acquired the fabric images through ESRF (European Synchrotron Radiation Facility) without disturbing the yarn and fabric structure. The contours of the yarn in the woven fabrics were accurately measured. They confirmed the non-circular nature of yarn cross-sections and also the flattening of yarns at the warp-weft intersections. Two polynomial functions were used to describe the shape of the varied cross-section. The cross-section shape which was modeled by elliptical and polynomial curves was variable along the yarn path. Figure 6 is the 3D representation in this method. Unfortunately, Gong didn't give a real 3D picture to demonstrate his research.

2.4. Based on Bezier Curve

Similar to Liao's method, Zhang [14] proposed that the central line of the yarn can be constructed by Bezier curve. The cross-section of the yarn is circular. First, a serial of interpolating points on the central line of the yarn are calculated or given. The control points determining the Bezier curve (central line) are reversely calculated by the interpolating points. Then, applying the sweeping technique, the vertexes of the control grid that determine the Bezier surface are calculated. Different from quadrilateral facets technology, the surface composes of many triangles which ensure the 3 vertexes of any triangle are at the same plane. The triangulation of the Bezier surface will be conducted automatically by OpenGL. The images of 3D yarn can be rendered by Computer Graphics principle.

2.5. Based on Spline Curve and Super Ellipse

Jiang and Chen [15] proposed an approach of super ellipse which can design the cross-section in arbitrarily shape. At the same time, the central line of the yarn is represented as spline curve that can pass the given control points. The shape of the cross-section can be designed manually, which makes it convenient to change the yarn shape. The uneven yarn can be modeled and bent. The algorithm can be used in constructing the fabric structure. The software Tex Gen is based on Chen's work.

2.6. Based on B-Spline Curve

Nowadays, using B-spline surface is the most popular method to model the configuration of 3D yarn. B-spline curve technology gives the designer more convenience to model the shape of the yarn accurately. Lin and Newton [16] used the cubic B-spline curve to generate the yarn paths, and circle as the form of cross-section. Recently, and Sherburn [17] are able to design irregular yarn using B-spline curve technology. Like Bezier

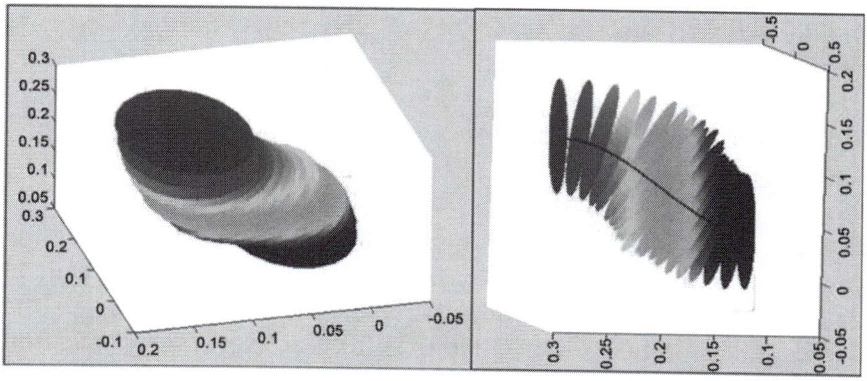

Figure 6. Gong's yarn representation in plain woven fabric [11] .

curve, the control points have to be calculated reversely by interpolating points. Unlike Bezier curve, the control points are dependent on the initial conditions, which also causes no uniqueness in control the shape of the yarn. If there is a tiny difference in initial conditions, for example, the positions of the end points change, there may be a big difference in the shape of the curve generated.

Zheng [18] [19] proposed an algorithm that the yarn can be modeled by NURBS (Non Uniform Rational B-Spline Curve) to keep a stable shape of the yarn. The shape preserve quasi uniform cubic B-spline curve is applied to fit the contour of the yarn path accurately, smoothly and stably. Two more adjacent points are inserted before and after a interpolating point respectively according to the lending direction of the all interpolating points, which ensures that the curve go through all the interpolating points. Therefore, the complicated reversely calculation is avoided in determining the control points. In Zheng's research, the quasi uniform quadratic B-spline curve is applied to approximately design the cross-section of a single yarn. The shape of the cross-section can be controlled by a 16-polygon formed by 18 control points. The shape of circle, ellipse, racetrack, lens, bowl, round rectangle can be designed along the yarn path simultaneously (see Figure 7). As long as the base cross-sections at given position are designed, the other cross-sections will be calculated automatically and smoothly. Each cross-section will be put perpendicular to central path. OpenGL is used to design NURBS surface and automatically generate the triangle facetsand give its shading effect. Moreover, some fancy yarns can also be designed by using different cross-section along the path, which will be introduced in the latter part.

Arif Kurbak [20] [21] also built 3D yarn model with 3DS-MAX for knitted fabric. The yarn paths were calculated according to the knitted structure. The cross-section of the yarn is set as circle, therefore, the yarn is actually a cylinder. The loop and tuck heads were taken as ellipses in two dimensions while the rest of the loops and tucks were taken as parabolic helices wrapped on circular cylinders. Thus, the path of yarn in different kind of knitted fabric can be clearly revealed.

To sum up, Table 1 gives a brief summary of characteristics of all the models introduced above.

3. SENSE OF REALITY

Although realistic rendering technology provided by OpenGL can draw a yarn with shade effect as shown in Figure 8, it is far from true yarn yet. As it is known to all that the yarn has the twist and hairiness.

Here are some woven models from famous fabric CAD company such as Scot Weave [22] , EAT [23] and Tex Gen [24] in Figure 9.

3.1. Twist

Zheng [25] added 1D and 2D texture on the surface of yarn to simulate the twist of yarn. For 1D texture as shown in Figure 10(A), parallel lines are drawn on the surface of the yarn with the help of OpenGL. The slop of the lines is dependent on the twist angle of the yarn. The lines are uniformly distributed and the twist effect will be worsen when magnify the yarn. For 2D texture mapping technology as shown in Figure 10(B), an image or a picture scanned from the real yarn will be mapped on the surface of the 3D model to reflect the twist and even hairiness. Ma [26] (Figure 10(C)) and NedGraphics Jacquard Software use the similar mapping technology. Obviously, it requires numerous pictures to imitate different twist effects for actual application.

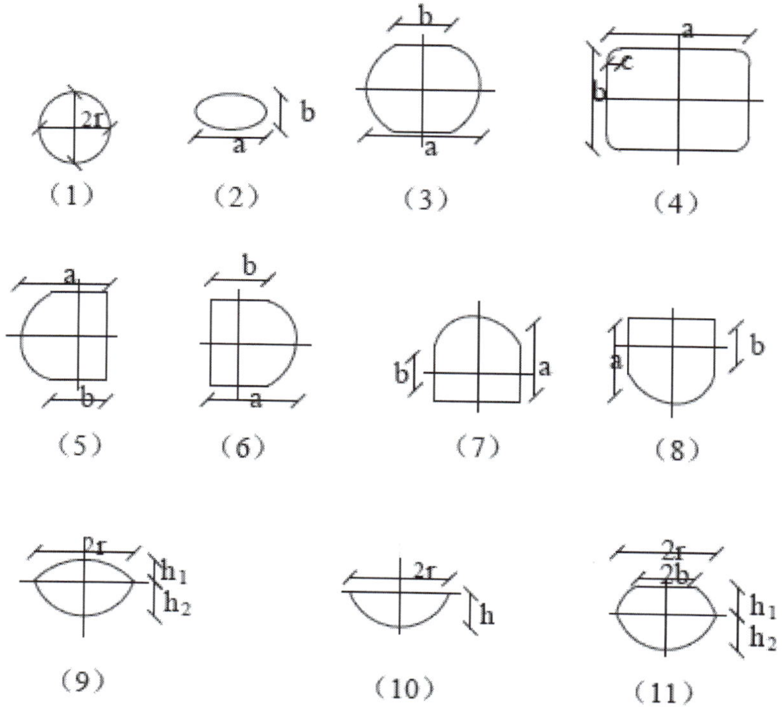

Figure 7. Cross-sections designable in Zheng's approach [18] .

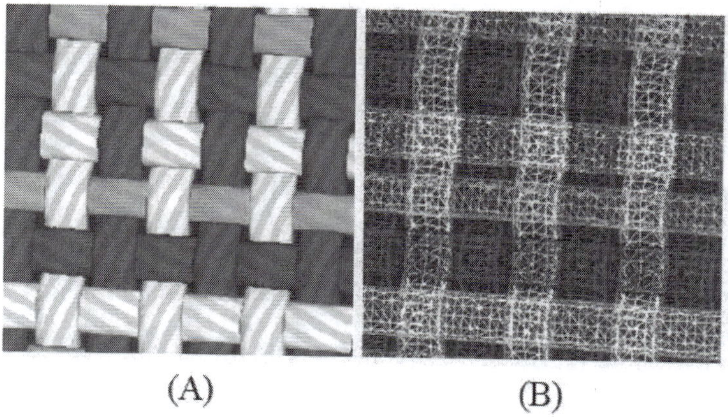

Figure 8. (A) 3D plain woven structure with 1D twist effect; (B) The triangle facets auto-generated by OpenGL.

Table 1. The characteristics of the various methods.

Proposer	Method	Cross-section	Central line	Characteristics	Implementing platform
Liao, Adanur	Sweeping the cross-section along the yarn path	Uniform ellipse, circle or race-track	Line segment and arc	Auto-generated 3D yarn, flexible	Unknown
Li	Assembling of circles	Circles of different diameters	Straight line	Straight yarn	MSVB
Lomov	Assembling of solid	circle	Calculated by WisTex	Uniform cylinder, flexible	MSVC
Lin, Long	Assembling of a serials of individual sections	Various cross-sections of quadratic B-spline curve, auto-generated by parameters	Bezier curve, calculate by TexGen	Shape and size of the cross-sections can be changed locally in fabric structure to avoid yarn penetrating	Python
Gong, Azgen	3D scanning of yarn by ESRF	Polynomial curve	Line segment and arc	The cross-sections are separated	MSVC
Zhang	Beizer suface	Beizer curve	Cubic Beizer curve	Uniform cylinder	MSVB,
Jiang, Chen	Spline curve and super ellipse	Uniform Super ellipse	Cubic spline curve, calculate by TexEng	The cross-section of the yarn can be edited,	Borland C,
Lin, Sherburn	B-spline surface	Various cross-sections	Cubic B-Spline curve	Auto-generated 3D yarn including fancy yarn	MSVC
Zheng	B-spline surface or NURBS	Various cross-sections of quadratic B-spline curve, auto-generated by parameters	Cubic shape-preserving B-spline curve	Auto-generated 3D yarn including fancy yarn	MSVC, OpenGL
Zheng, Zhao	Assembling of a serials of individual sections	C-cardinal spline curve, arbitrary shape,	C-cardinal spline curve	Cross-section of the yarn including fancy yarn can be edited manually	MSVC, OpenGL
Arif Kurbak	Circle and parabolic helix	Circle	Ellipse or parabolic helix in 2 dimensions	Circular cylinder	3DS-MAX

In order to improve the twist effects, Zheng [19] [27] [28] added bumping texture to acquire more realistic effect or 3D twist effect. By adding troughs and ridges on the contour of the cross-section of the yarn, the surface of the 3D yarn will appear bright and shade effect. When transforming the troughs and ridges at different sections, the twist effect will appear. If a trough or a ridge is deliberately discarded at random, the migration effect of fibers in the structure of the yarn is imitated. The principle is explained in Figure 11 and the imitated image is shown in Figure 12.

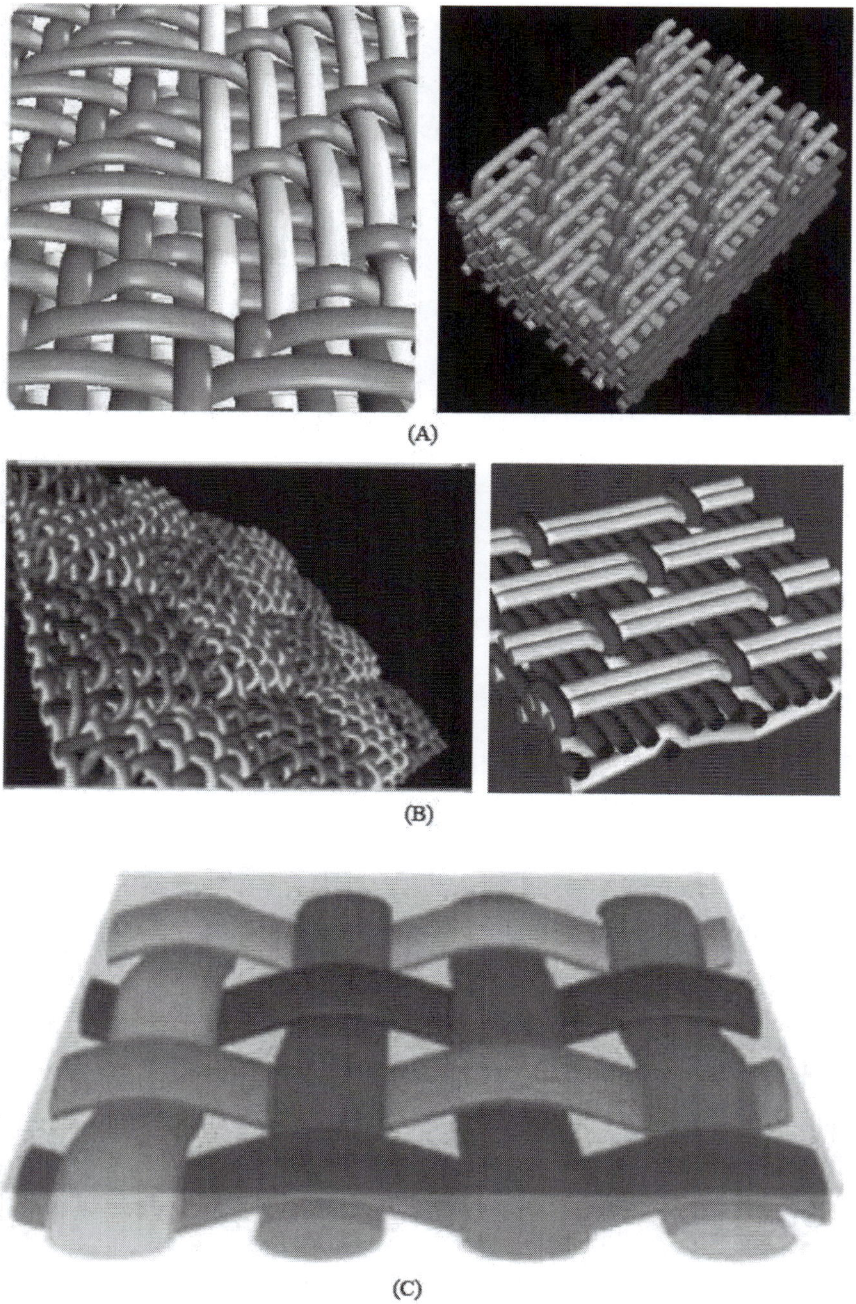

Figure 9. Fabric Simulation by some companies: (A) Scot Weave [22] ; (B) EAT [23] and (C) TexGen [24] .

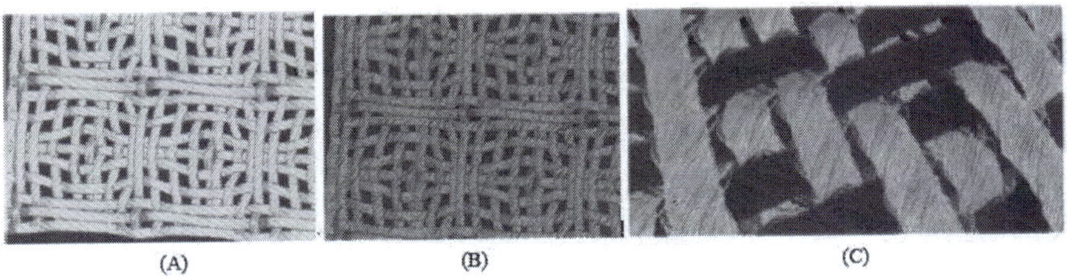

Figure 10. Zheng's fabric simulating with twist effect (A) 1D texture for twist; (B) 2D texture mapping for twist [25] and (C) is Ma's study of weave geometries incorporating twisted fiber bundles [26] .

Zhao [29] proposed another algorithm to model the yarn based on slice superposition method and C-Cardinal spline curve, which is similar to the shape preserve B-spline curve. The curve passes through the interpolating points except the end points. The drawback can be avoided by repeating the end points two times. Similar to the bumping texture in Zheng [19] , the cross-section of the yarn is an irregular polygon as shown in Figure 13(A). By rotating the polygons, a number of polygons are obtained and the series of polygon slices are combined to form a yarn with 3D twist effect as shown in Figure 13(B). The advantages of using the C-Cardinal spline curve are embodied in easy calculation, small memory required. However, the disadvantage is that the curve has not been supported by OpenGL or other software package. If the 3D image is enlarged enough, the slices are separated and the yarn appears discontinuously.

Sriprateep [30] also proposed an algorithm based on fibers' migration in yarns to imitate the 3D twist effect of the ring spun yarn. Each yarn is considered as the assembly of fibers. Each fiber is regarded as a helix along yarn path. The cross-section of the yarn is thought as elliptical shape. Therefore, the yarn can be approximately

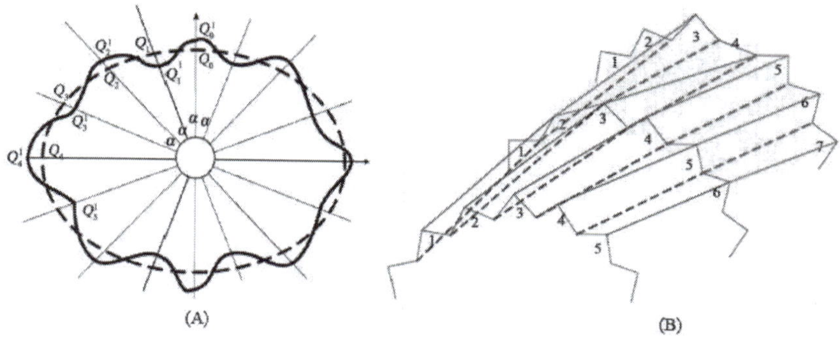

Figure 11. Zheng added bumping texture as (A) and then (B) the trough and ridge in each cross-section is discarded randomly to create the migration effect of the fiber [19] .

Figure 12. Zheng's yarn model with 3D twist effect in this method [28] .

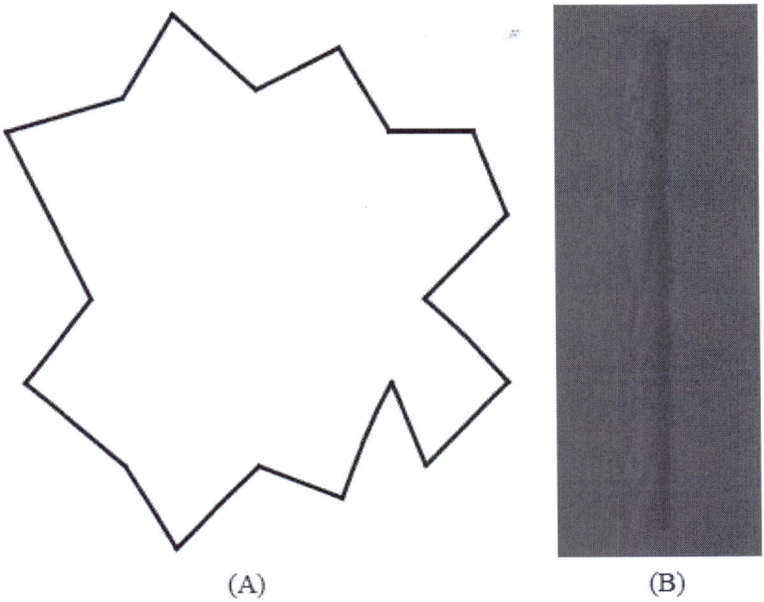

Figure 13. Zhao using an irregular polygon as cross-section as (A) and (B) is the 3D yarn model with twist effect [29] .

constructed by NURBS. The migration parameters include twisting tension, migration period and amplitude, initial phase and frequency of migration pattern. The geometry of yarn structures is eventually modeled by using the Solid Works CAD software package and the model is validated by imitating the process of spinning rayon yarn (see Figure 14). The difference between the algorithm and Zheng's is that Scriprateep uses many NURBS surfaces while Zheng only uses single NUBRS surface to imitate 3D twist effect.

3.2. Hairiness

A realistic model for 3D yarn should reflect the hairiness effect. Zhong [31] achieved the aim by using projected triangles to simulate the hairiness, but the effect could be weakened if the 3D image was enlarged. As Figure 15(A) shown, Xu [32] realized a photorealistic effect yarn by rotating an image including many lighten points. The shade effect, twist effect and hairiness effect were demonstrated excellently. However, it is not a real 3D model for yarn as the image generated cannot be rotated or viewed in different direction. Effect in this approach is in Figure 15(B). Zheng [19] proposed that NURBS surface be used to imitate the projected fiber similar to the technique for yarn. The length, projecting angle and the distribution of the hairs can be designed.

4. APPLICATION OF 3D SINGLE YARN

4.1. Folded Yarn

Folded yarn comprises of two or more single yarn, so it can be designed if the central paths of all the component yarns are determined. In Zheng [33] , a two-ply or three-ply folded yarn is modeled by rotating two or three circular cross-section around so called folded yarn center, therefore, the central line of each component single yarn is actually a helix. The sizes of the twist are controlled by the screw pitch of the helix while the direction of the twist is determined by the rotating angle. In Zhao and Zheng [34] , the cross-section of the single yarn in the folded yarn was assumed to be circular or irregular polygon by slice superposition method which is shown inFigure 16(A). In order to demonstrate the twist effect of the folded yarn, at least 4 slices of cross-sections and their centers of each component yarn will be calculated for every other 90°. Figure 16(B) shows a bent folded yarn based on the algorithm similar to Zheng [22] [28] . If more than 20 single yarns are folded together, the folded yarn appears as a single yarn as each yarn is so slim that looks like a single fiber. Figure 16(C) gives the effects which is similar to Sriprateep's work [30] .

4.2. Fancy Yarn

It is certain that the fancy yarn can be modeled by 3D technology although it is rare. For fancy yarn, there are varied sections along the yarn path. It can be folded yarn whose components are of various color, thickness or configuration in space. Based on Zheng's [19] [35] [36] algorithm, the cross-sections and configuration of the yarn can be changed based on NURBS, the fancy yarn will be modeled. Figure 17(A) demonstrates a slub yarn

Figure 14. Sriprateep's yarn model with 3D twist effect which based on fiber's migration [30] .

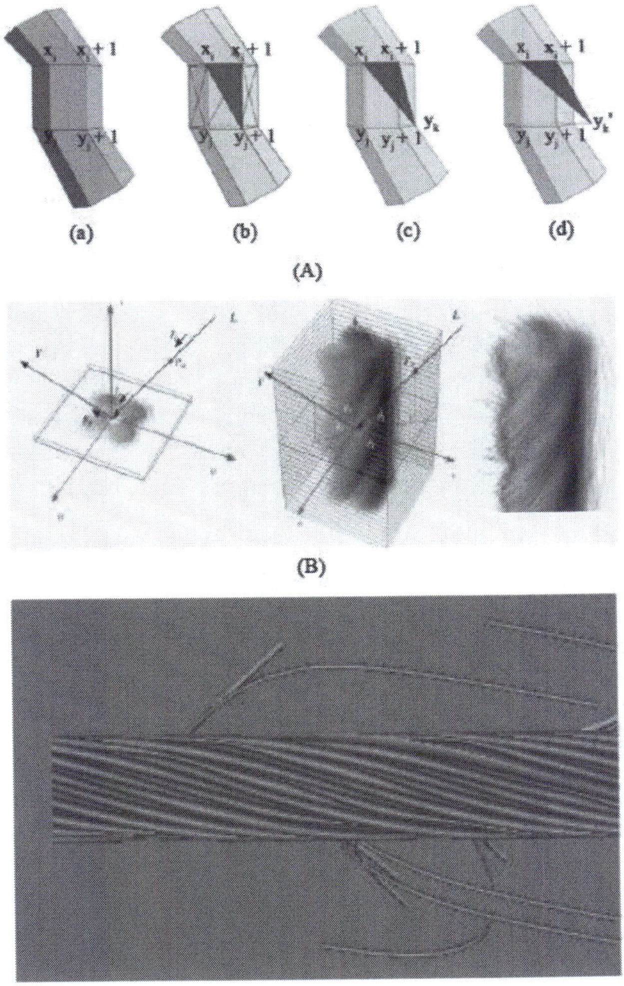

Figure 15. Hairiness effect in different methods: (A) is Zhong's yarn mo- del with hairiness built by projected triangles [31] ; (B) Xu's yarn model with photorealistic effect based on image [32] ; (C) Zheng added hairiness by NURBS surface [19] .

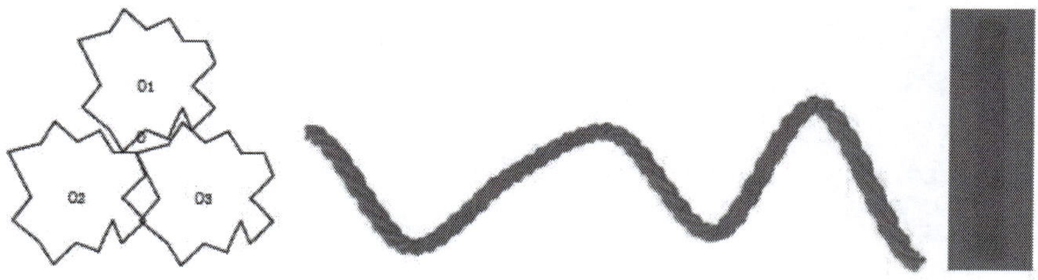

Figure 16. (A) Irregular polygon in Zhao's yarn model; (B) Zhao's 3D model of folded yarn; (C) A single yarn is imitated by modeling multi-folded yarn [34] .

modeled on the principle. In fact, loop yarn, knop yarn, slub yarn, snarly yarn can also be designed when the frequency/size of the effect spot, path/color of the component single yarn are designed. Based on C-Cardinal spline curve, Zhao [34] demonstrates a slub yarn as shown as Figure 17(B). The slub yarn includes even parts and uneven parts which change from thinner to the thicker diameter. The twist effect can be improved by increasing the rotating a larger angle of the two neighboring slices at the thinner area.

Zhuge [36] designed the contour of fancy yarn was drawn by Bezier tools. Then template of illumination factor was applied to the model. Then, loop hairiness, end hairiness or floating hairiness can be generated by random noise point.

4.3. Structure of Fabric

If the geometric structure in a woven, knitted or braid fabric is calculated, or in other words, the shape and the path of the each yarn in fabric is determined, the 3D geometric structure of the fabric can be imitated. Figure 18(A) shows a plain fabric with the diameter of the yarn changes along the central path; Figure 18(B) shows the same fabric structure with 3D twist effect; Figure 18(C) shows a plain fabric comprised of folded yarn; Figure 19(A) and Figure 19(B) [35] show two fancy woven fabrics respectively; Figure 20 shows a knitted fabric structure.

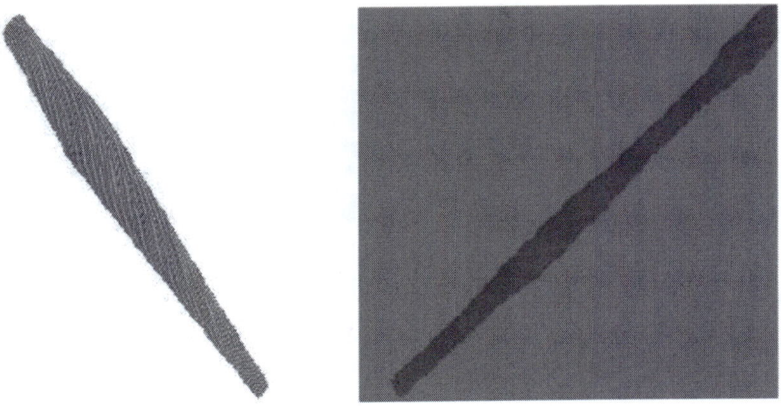

(A) (B)

Figure 17. (A) Zheng's yarn model of section part in slub yarn [19] ; (B) Zhao's slub yarn model with different twist in the thicker and thinner area [34] .

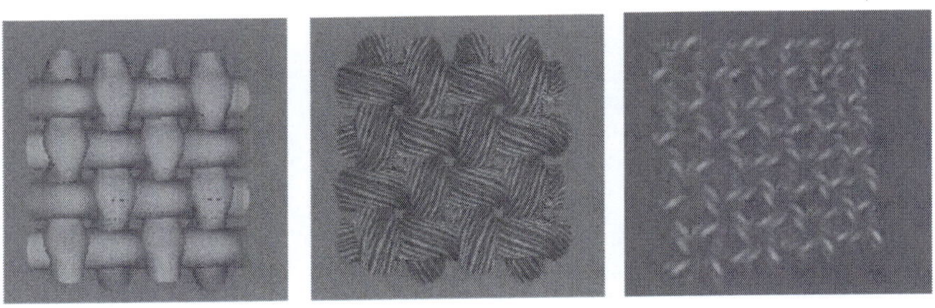

(A) (B) (C)

Figure 18. (A) Plain fabric of various cross-sections without twist; (B) Plain fabric of various cross-sections with twist; (C) Plain fabric comprised of fold- ed yarn.

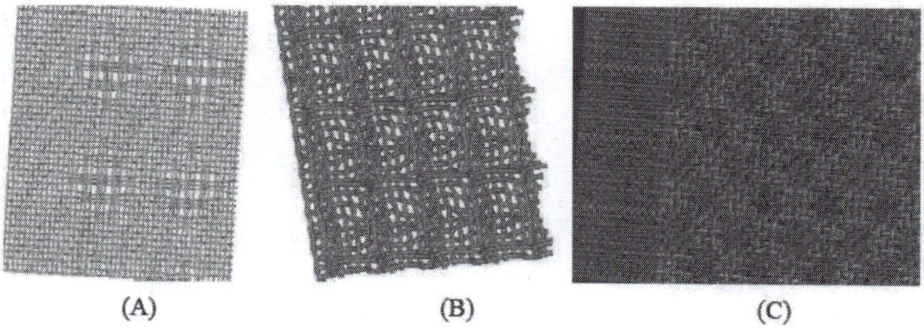

Figure 19. Structure of fancy weaves (A) Effect of mesh weaves; (B) Honey comb weaves [36] ; (C) The structure of interchanging double cloth.

Figure 20. Structure of knitted fabric in Kayacan's study [37] .

5. CONCLUSION

Multiple approaches have been reviewed for modeling 3D yarn geometric configuration. The yarns are shaped on the theory of Peirce while the 3D constructing technology is based on Computer Graphics. And all the methods are related with designing the cross-sections along the yarn path. Comparatively speaking, the method based on spline curve is the most popular one since it gives the most flexibility in designing not only non-uniform single yarn with different cross-sections and twist effect, but also folded yarn, fancy yarn including slub yarn, knep yarn. The yarn can bent or deform in 3D space freely so it can be used in woven fabric's structure conveniently. The model can also be applied to the structure of the knitted fabric, braided fabric so long as the central line of the yarn is given.

ACKNOWLEDGEMENTS

Tianyong Zheng, Jing Wei, Zhengtao Shi, Tingting Li, Zhen Wu Authors would like to thank Henan Provincial Collaborative Innovation Center for Textiles and Appeals, Zheng- zhou Municipal Bureau of Science and Technology for supporting the research.

REFERENCES

1. Hearn, D. and Baker, M.P. (1998) Computer Graphics (C Version). Prentice Hall, Inc., Upper Saddle River, 335-355.
2. Pigel, L. and Tiller, W. (1996) The NURBS Book. 2nd Edition, Springer Press, Berlin.
3. Angel, E. (2003) Interactive Computer Graphics: A Top-Down Approach with OpenGL. 3rd Edition, Person, Addison Wisley, Boston, 193-194.
4. Peirce, F.T. (1937) Geometry of Cloth Structure. Textile Research Journal, 28, 45-96.
5. Kemp, A. (1958) An Extension of Peirce's Applicable to the Treatment of Non-circular Threads. Journal of the Textle Institute, 49, 45.
6. Hearle, J.W.S. and Shanahan, W.J. (1978) An Energy Method for Calculations in Fabric Mechanics. Part I: Principles of the Method. Journal of the Textle Institute, 69, 81-89.

7. Shanahan, W.J. and Hearle, J.W.S. (1978) An Energy Method for Calculations in Fabric Mechanics. Part II: Examples of Application of the Method to Woven Fabrics. Journal of the Textle Institute, 69, 81-89.

8. Liao, T. and Adanur, S. (1998) A Novel Approach to Three-Dimensional Modeling of Interlaced Fabric Structures. Textile Research Journal, 68, 841-847.

9. Lomov, S.V., Mikolanda, T., Kosek, M. and Verposest, I. (2006) Model of Internal Geometry of Textile Fabrics: Data Structure and Virtual Reality Implementation. Journal of the Textile Institute, 98, 1-13.

10. Li, Y., Zeng, P. and Zhao, L. (2004) The Study on the Clothing Simulation Based on 3D Graphics. Journal of Textile Research (China), 25, 67-69.

11. Lin, H., Zeng, X., Martin, S., Andrew, L. and Mike, C. (2011) Automated Geometric Modelling of Textile Structures. Textile Research Journal, 82, 1689-1702.

12. Gong, R.H., Ozgen, B. and Soleimani, M. (2009) Modeling of Yarn Cross-Section in Plain Woven Fabric. Textile Research Journal, 79, 1014-1020.

13. Ozgen, B. and Gong, R.H. (2011) Modelling of Yarn Flattening in Woven Fabrics. Textile Research Journal, 81, 1523-1531.

14. Zhang, R., Huang, X. and Li, R. (2005) 3D Computer Simulating of Weave Fabric. Journal of Textile Research (China), 26, 42-44.

15. Jiang, Y. and Chen, X. (2004) The Structural Design of Software for Simulation and Analysis of Woven Fabric by Geometry Model. Textile Institute World Conference (China), Donghua University Press, Shanghai, 1298-1301.

16. Lin, H.Y. and Newton, A. (1999) Computer Representation of Woven Fabric by Using B-Splines. Journal of the Textile Institute, 90, 59-72

17. Sherburn, M. (2007) Geometric and Mechanical Modeling of Textiles. Ph.D. Thesis, Thesis Submitted to The University of Nottingham for the Degree of Doctor of Philosophy.

18. Zheng, T.Y. and Cui, S. (2006) The Study on Constructing the 3D Yarn Model by B-Spline Surface. Part I: Construction of 3D Yarn Model with Different Cross Sections. Journal of Textile Research (China), 27, 53-57.

19. Zheng, T.Y. and Cui, S. (2007) Improvement on Modeling a 3D Yarn with B-Spline Surface. Journal of Textile Research, 28, 35-44.

20. Kurbak, A. and Soydan, A.S. (2009) Geometrical Models for Balanced Rid Knitted Fabrics Part III: 2 × 2, 3 × 3, 4 × 4, and 5 × 5 Rib Fabrics. Textile Research Journal, 79, 618-625.

21. Kurbak, A. and Soydan, A.S. (2009) Geometrical Models for Cardigan Structures. Part I: Full Cardigan. Textile Research Journal, 79, 1281-1300.

22. http://scotweave.com/products/technical-weaver/

23. http://www.designscopecompany.com/cross-section/

24. http://www.bluefox.com

25. Zheng, T.Y. and Huang, G. (2002) The Setup of 3D Model for Fabric Yarns (China). Journal of Textile Research, 23, 13-15.

26. Ma, L., George, B., Hu, J. and Zhang, J. (2011) A Novel Weave Pattern Encoding Method Using Neighbor Information and Its Applications. Textile Research Journal, 81, 632-648.

27. Zheng, T. and Cui, S. (2006) The Study on Constructing the 3D Yarn Model by B-Spline Surface. Part II: Inserting 3D Bumping Texture on the Surface of the Yarn. Journal of Textile Research (China), 27, 53-57.

28. Zheng, T.Y. (2010) Study on General Geometrical Modeling of Single Yarn with 3D Twist Effect. Textile Research Journal, 80, 867-879.

29. Zhao, Z., Zheng, T.Y., Wu, Z. and Li, T. (2014) Study on Constructing the Model of Ply Yarn by the Cubic C-Cardinal Spline Cure. Journal of Textile Research, in press.

30. Sriprateep, K. and Bohez, E.L.J. (2009) A New Computer Geometric Modelling Approach of Yarn Structures for the Conventional Ring Spinning Process. Journal of Textile Institute, 100, 223-236.

31. Zhong, H., Xu, Y., Guo, B. and Shum, H. (2001) Realistic and Efficient Rendering of Free-Form Knitwear. The Journal of Visualization and Computer Animation (China), 12, 13-22.

32. Xu, Y. (2000) Photorealistic Rendering of Knitwear Using the Lumislice. Institute of Software, Academia China.

33. Zheng, T.Y. (2001) A 3D Model for Folded Yarns. Journal of Textile Research (China), 23, 20-21.

34. Zhao, Z. (2014) Three-D Simulation Research of Ply Yarn Based on the Cubic C-Cardinal Spline Cure Modeling Techniques. Master's Thesis, Zhongyuan University of Technology, Zhengzhou.

35. Zheng, T.Y. (2005) 3D Appearance Imitation of Fancy Weaves. Journal of Textile Research (China), 26, 35-38.

36. Zhuge, Z. and Zhang, F. (2005) Numerical Simulation of Fancy Yarn. Journal of Zhejiang University (China), 39, 1529-1536.

37. Kayacan, O. and Kurbak, A. (2008) Basic Studies for Modeling Complex Weft Knitted Fabric Structures Part IV: Geometrical Modeling of Miss Stitches. Textile Research Journal, 78, 659-663.

13

CHAPTER

Mathematical Modelling of Dry Crease Recover Angle for Single Step Fixation of Reactive Printing and Crease Resistance Finishing of Cotton Fabric Using DOE

Fareha Asim[1], Muzzaffar Mahmood[2]

[1]*Department of Textile Engineering, NED University of Engineering & Technology Karachi, Pakistan*

[2]*Department of Mechanical Engineering, NED University of Engineering & Technology Karachi, Pakistan*

ABSTRACT

The study was carried out to develop the mathematical model of crease recovery angle for single step fixation process of cotton fabric for reactive printing and crease resistance finishing using design of experiment.Single step fixation process using Monochlorotriazine (MCT) based reactive dye and modified dimethyloldihydroxy ethylene urea (DMDHEU) based crease resistantwas investigated to model the relationship of significant factors on the dry crease recovery angle. The effects of several factors governing the single step process, such as dye concentration, concentration of crease resistant, fixation method and temperature, on dry crease recovery angle were examined.Initially sets of screening experiments have been conducted to identify the significant factors and their interaction.Furthermore, the detailed analysis of significant factors including ANOVA, residual analysis, model adequacy checking and regression analysis has been performed.However, mathematical model have been developed as a function ofsignificant factors namely concentration of crease resistant, fixation method and fixation temperature. The statistical analysis of the significant factors revealed the quadratic effect of each factor on the crease recovery anglein single step fixation process. The uncertainty analyses for measurements shows that the predicted values are in good agreement with experimental data and are sufficiently accurate.

Keywords: Crease Resistance Finishing, Design of Experiment, Dry Crease Recovery Angle, Mathematical Modelling, Reactive Printing

1. INTRODUCTION

In Textile processes, it is often of primary interest to develop the relationships between the key process variables and the performance characteristics. MathematicalModelling play an important role in developing, analysing and predictingthe relationship from experiments with certain level of confidence[1-6].When several variables influence a certain characteristic of a product, the best strategy is then to design an experiment(DOE) so that objective conclusions can be drawn rather than judgemental. In the context of DOE in Textiletwo types of process variables: qualitative and quantitative factors[7-12]. The aim of this work was to develop the mathematical model of crease recovery angle for single step fixation process of reactive printing and crease resistance finishing process.The significant factors for the single step fixation of reactive printing and crease resistance finishing process had been reviewed previously [13-16].

In the current studyof single step fixation process of reactive printing and crease resistance finishing chroma, concentration of crease resistant and fixation temperature are quantitative factors. However fixation method is the qualitative factor.

2. EXPERIMENTAL

2.1. Process Route

The singlestep fixation of reactive printing and crease resistance finishing is carried outas follows: Crease Resistant Finish-Dry-Reactive Print-Fix-Wash-Dry-Conditioned.

In the first step ready to print fabric was padded with thecrease resistant finish liquor, and then squeezed to obtain a 70% wet pickup. The fabric was then dried at 100°C for 1 min. The dried fabric was printed by lab scale Rotary Printing machine. The finish-print fabric was concurrently fixed using Econtrol method at 135°C for 5 minutes. The fixed samples were finally washed, dried and conditioned.

2.2. Evaluation of Dry Crease Recovery Angle (DCRA)

The crease resistant properties were evaluated by measuring the dry crease recovery angles (DCRA). The print-finish fabrics were conditioned (at temperature 25±1°C and relative humidity 65 ±1%) before the measurement of DCRA values. The dry crease recovery angle (DCRA) of the fabric was measured using AATCC test method 66-1990. The wider the DCRA is, the higher the crease recovery. The samples processed with single step fixation were compared with those produced from a conventional two-step process of reactive printing and Crease resistance finishing.The experiment data is shown in Table 1 and 2.

2.3. Statistical Analysis of the Data

After processing the experimental trials, DCRAvalues were measured using crease recovery tester.The results obtained from the different experimental sets are presented in Table 2. The DCR Avalues were statistically analysed. Residual analysis and model adequacy checking were performed. The experiments have been conducted in random order. The results were analysed using software Design Expert 8.0.

2.4. Model Adequacy Checking

2.4.1. The Normality Assumption

The normal probability plot of the residual for dry crease recovery angles has been shown in Figure 1. There is no severe indication of non-normality, nor is there any evidence pointing to possible outliers and the equality of variance assumption does not seem to be violated.

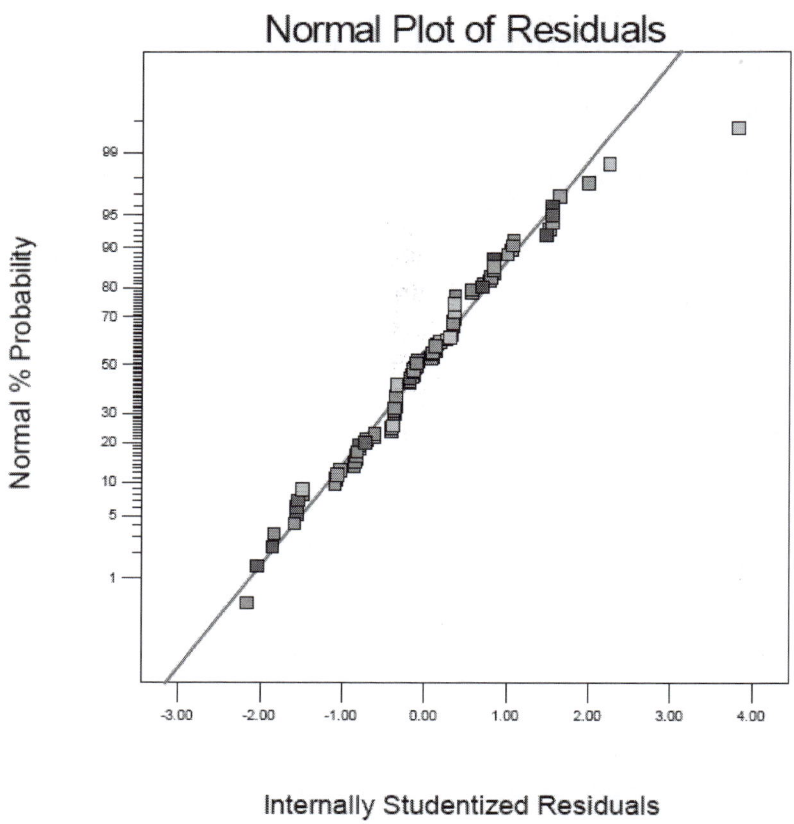

Figure 1. Normal plot of residuals for DCRA

2.4.2. Residual Analysis

Rresidualversus run number and residual versus predicted plots of dry crease recovery angle have been shown from Figure 2and 3. The pattern of the residualsshowed that they are normally distributed and the equality of variance does not seem to be violated.

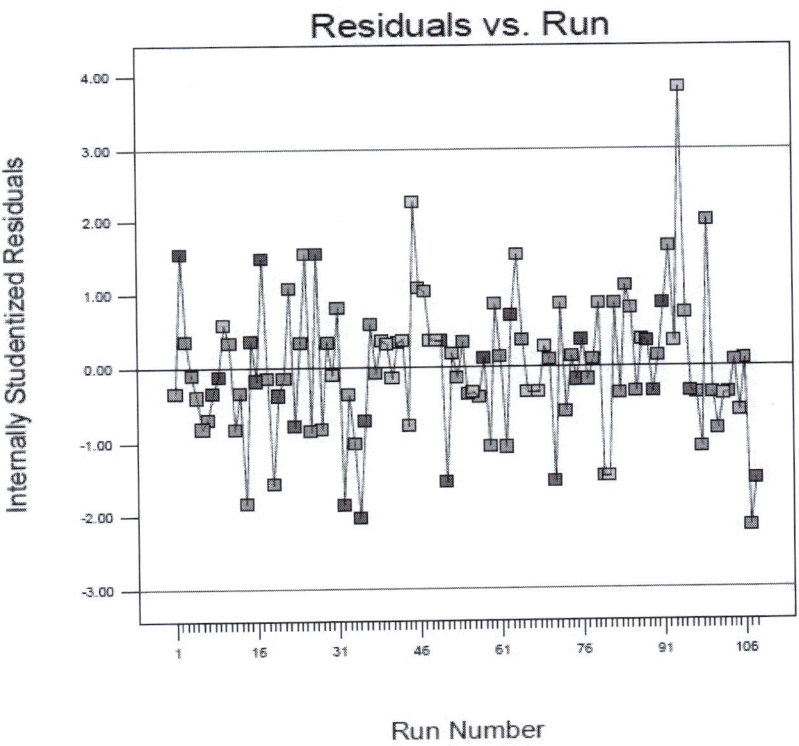

Figure 2. Residual vs. run number for DCRA

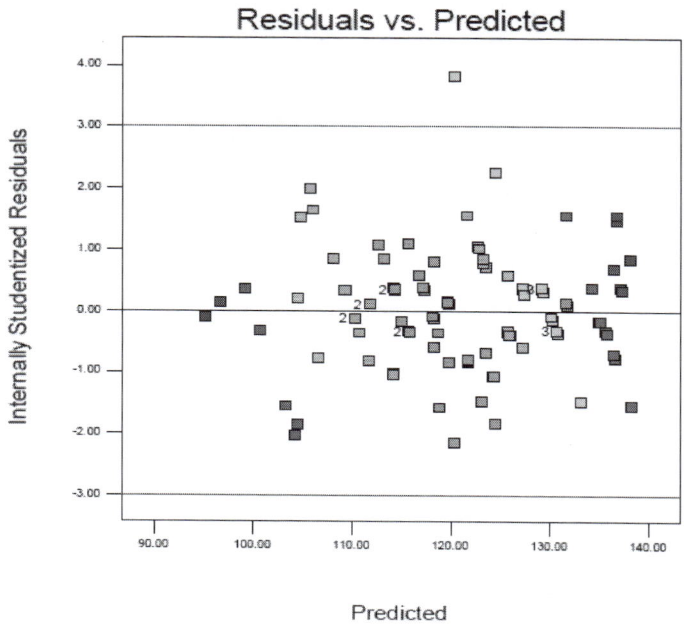

Figure 3. Residual vs. predicted plot for DCRA

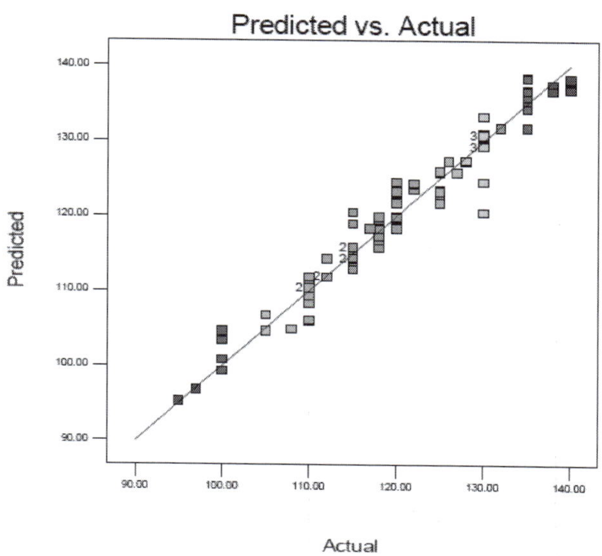

Figure 4. The predicted vs. actual plot for DCRA

2.5. Analysis of Variances (ANOVA)

Table 1 shows the summary of ANOVA results ofdry crease recovery angle for single step fixation process of reactive printing and crease resistance finishing. The model p-value implied that the respective model is significant for dry crease recovery angle.As can be seen in Table 1,

The "Lack of Fit p-value" of 0.9998 implies the Lack of Fit is not significant relative to the pure error representing that the suggested models are well-fitted to the observed response.. The "Pred R-Squared" of 0.8630 is in reasonable agreement with the "Adj R-Squared" of 0.9189 specifying the respective model is adequate.

The actual and predicted plot of dry crease recovery angle has been shown in Figure 4. The actual values are the measured response data for particular run and the predicted values are theresults generated using the approximate functions. It is found that the adequate precision of

Dry crease recovery angle is found to be 20.27, which measured the signal to noise ratio is much greater than 4, (Table 1). This indicates the modelis adequate to be used to navigate the design space. Focusing on the model maximizing the "Adjusted R-squared" and the "Predicted R-squared" different process order selection has been made for dry crease recovery angle. The alias structures of the dry crease recovery angle are shown in Table 3.

Table 1. Summary of ANOVA results of dry crease recovery angle (DCRA)

Source	Sum of Squares	Degree of freedom	Mean Square	F-value	p-value	Remarks
DCRA Warp (°): Sixth Model						
Model	11315.30	52	217.60	24.10	< 0.0001	**Significant**
Residual	487.53	54	9.03			
Lack of fit	303.53	49	6.19	0.17	0.9998	**Not significant**
Pure error	184.00	5	36.80			
R^2 = 0.9587,Pred R^2 0.8630, R^2_{adj} = 0.9189; Adequate Precision = 20.279						

Table 2. ANOVA for DCRA warp

Source	Sum of Squares	Degree of freedom	Mean Square	F-value	p-value
Block	164.03	1	164.03		
Model	11315.30	52	217.60	24.10	< 0.0001
B-Concentration of CR	666.13	1	666.13	73.78	< 0.0001
C-Fixation Method	77.68	1	77.68	8.60	0.0049
D-Fixation Temperature	220.50	1	220.50	24.42	< 0.0001
AC	134.84	1	134.84	14.94	0.0003
AD	522.58	1	522.58	57.88	< 0.0001
BC	276.13	1	276.13	30.58	< 0.0001
B^2	82.38	1	82.38	9.12	0.0038
ABC	70.95	1	70.95	7.86	0.0070
ABD	172.14	1	172.14	19.07	< 0.0001
ACD	106.84	1	106.84	11.83	0.0011
BCD	115.56	1	115.56	12.80	0.0007
A^2B	364.19	1	364.19	40.34	< 0.0001
A^2C	363.01	1	363.01	40.21	< 0.0001
AD^2	115.12	1	115.12	12.75	0.0008
B^2C	441.83	1	441.83	48.94	< 0.0001
B^2D	63.02	1	63.02	6.98	0.0108
BD^2	247.52	1	247.52	27.42	< 0.0001
CD^2	72.21	1	72.21	8.00	0.0066
ABCD	45.54	1	45.54	5.04	0.0288
A^2B^2	77.11	1	77.11	8.54	0.0051
A^2BC	54.91	1	54.91	6.08	0.0169
A^2BD	38.53	1	38.53	4.27	0.0437
A^2CD	56.08	1	56.08	6.21	0.0158
A^2D^2	78.96	1	78.96	8.75	0.0046
AB^2C	44.04	1	44.04	4.88	0.0315
AB^2D	91.18	1	91.18	10.10	0.0025
ACD^2	78.26	1	78.26	8.67	0.0048
B^2D^2	333.86	1	333.86	36.98	< 0.0001
BCD^2	475.02	1	475.02	52.61	< 0.0001
A^2B^2C	148.14	1	148.14	16.41	0.0002
A^2B^2D	62.04	1	62.04	6.87	0.0114
A^2BD^2	151.34	1	151.34	16.76	0.0001
A^2CD^2	108.69	1	108.69	12.04	0.0010
AB^2CD	181.16	1	181.16	20.07	< 0.0001
AB^2D^2	55.46	1	55.46	6.14	0.0163
$ABCD^2$	176.49	1	176.49	19.55	< 0.0001
B^2CD^2	125.14	1	125.14	13.86	0.0005
$A^2B^2D^2$	269.95	1	269.95	29.90	< 0.0001
A^2BCD^2	69.79	1	69.79	7.73	0.0075
AB^2CD^2	73.99	1	73.99	8.20	0.0060
Residual	487.53	54	9.03		
Lack of Fit	303.53	49	6.19	0.17	0.9998
Pure Error	184.00	5	36.80		
Cor Total	11966.85	107			

Std. Dev.	3.00	C.V%	2.47	PRESS	1616.45
R^2	0.9587	Adj R^2	0.9189	Pred R^2	0.8630

Table 3. Alias structures for DCRA

S.No.	Aliases
1.	$[A] = A^3 = A^5$
2.	$[B] = B^3 = B^5$
3.	$[D] = D^3 = D^5$
4.	$[AB] = A^3B = AB^3 = A^3B^3 = A^5B = AB^5$
5.	$[AC] = A^3C = A^5C$
6.	$[AD] = A^3D = AD^3 = A^3D^3 = A^5D = AD^5$
7.	$[BC] = B^3C = B^5C$
8.	$[BD] = B^3D = BD^3 = B^3D^3 = B^5D = BD^5$
9.	$[CD] = CD^3 = CD^5$
10	$[A^2] = A^4 = A^6$
11.	$[B^2] = B^4 = B^6$
12.	$[D^2] = D^4 = D^6$
13.	$[AB^2] = A^3B^2 = AB^4$
14.	$[ABC] = A^3BC = AB^3C$
15.	$[ABD] = A^3BD = AB^3D = ABD^3$
16.	$[ACD] = A^3CD = ACD^3$
17.	$[AD^2] = A^3D^2 = AD^4$
18.	$[A^2B] = A^2B^3 = A^4B$
19.	$[A^2D] = A^2D^3 = A^4D$
20.	$[BCD] = B^3CD = BCD^3$
21.	$[BD^2] = B^3D^2 = BD^4$
22.	$[B^2D] = B^2D^3 = B^4D$
23.	$[A^2C] = A^4C$
24.	$[B^2C] = B^4C$
25.	$[CD^2] = CD^4$
26.	$[AB^2C] = A^3B^2C = AB^4C$
27.	$[AB^2D] = A^3B^2D = AB^2D^3 = AB^4D$
28.	$[ABCD] = A^3BCD = AB^3CD = ABCD^3$
29.	$[ABD^2] = A^3BD^2 = AB^3D^2 = ABD^4$
30.	$[ACD^2] = A^3CD^2 = ACD^4$
31.	$[A^2BC] = A^2B^3C = A^4BC$
32.	$[A^2BD] = A^2B^3D = A^2BD^3 = A^4BD$
33.	$[A^2CD] = A^2CD^3 = A^4CD$
34.	$[BCD^2] = B^3CD^2 = BCD^4$
35.	$[B^2CD] = B^2CD^3 = B^4CD$
36.	$[A^2B^2] = A^4B^2 = A^2B^4$
37.	$[A^2D^2] = A^4D^2 = A^2D^4$
38.	$[B^2D^2] = B^4D^2 = B^2D^4$

3. RESULTS AND DISCUSSION

3.1. Analysis of Dry Crease Recovery Angle (DCRA)

The sixth order model has been selected for the response variable DCRA Warp. The detail of ANOVA of DCRA Warp is shown in Table 2. The Model p-value less than 0.05 implies the model is significant. In this case *B, C, D, AC, AD, BC, B2, ABC, ABD, ACD, BCD, A2B, A2C, AD2, B2C, B2D, BD2, CD2, ABCD, A2B2, A2BC, A2BD, A2CD, A2D2, AB2C, AB2D, ACD2, B2D2, BCD2, A2B2C, A2B2D, A2BD2, A2CD2, AB2CD, AB2D2, ABCD2, B2CD2, A2B2D2, A2BCD2, AB2CD2*are significant model terms.

The graphs of significant model terms for DCRA have been shown from Figure 5 to 15.The graphs of significant model terms revealed that the dry crease revovery angle of the single step fixation process of reactive printing and crease resistance finishing is dependent on the interaction and curvature effect of concentration of crease resistant, fixation mode and fixation temperature.Figure 8 and 9 indicated that the Econtrol process supersedes the conventional curing process for fixation and gives a higher dry crease recovery angle in the single step fixation process. Figure 16illustrates the interaction plot of fixation mode and fixation temperature. The graph clearly shown that the Econtrol process at 130°C gives a higher dry crease recovery angle in the CPF process. The reduced regression equation of dry crease recovery angle has been shown in Equation (1).Whereas, the fitted regression equation has been presented in Equation (2). The alias structures are shown in Table 3.

3.2. Mathematical Model with Significant Variables at α = 0.05

$$\begin{aligned} DCRAWarp = {} & 125.09+1.59A+1.90B+4.33C+3.66D-2.32AB-1.97AC \\ & +3.56AD+0.62BC-0.14BD-4.12CD-1.23A^2-0.67B^2-3.35D^2 \end{aligned} \quad (1)$$

3.3. Fitted Regression Equation of DCRA

$$\begin{aligned} DCRAWarp &= 123.15 - 0.83A + 9.12B - 3.49C + 5.25D - 0.82AB - 4.58AC + 5.60AD - 5.87BC - 1.44BD - \\ & 1.50CD + 0.68A^2 + 5.23B^2 + 2.85D^2 + 2.19ABC + 2.39ABD - 2.53ACD - 2.69BCD - 8.38A^2B + 8.65A^2C - \\ & 1.85A^2D + 0.90AB^2 + 4.99AD^2 + 9.44B^2C - 3.44B^2D + 3.82CD^2 + 1.24ABCD - 6.55A^2B^2 + 3.25A^2BC + 1.92 \\ & A^2BD - 3.22A^2CD - 6.59A^2D^2 + 3.14AB^2C - 2.92AB^2D - 1.79ABD^2 + 4.11ACD^2 - 2.44B^2CD - 13.16B^2D^2 \\ & + 9.44BCD^2 - 5.26A^2B^2C + 4.17A^2B^2D + 1.01A^2BCD + 6.61A^2BD^2 - 4.54A^2CD^2 + 4.11AB^2CD - 4.20AB^2D^2 \\ & + 4.23ABCD^2 - 4.99B^2CD^2 + 2.45A^2B^2CD + 15.04A^2B^2D^2 - 4.49A^2BCD^2 - 4.84AB^2CD^2 \end{aligned} \tag{2}$$

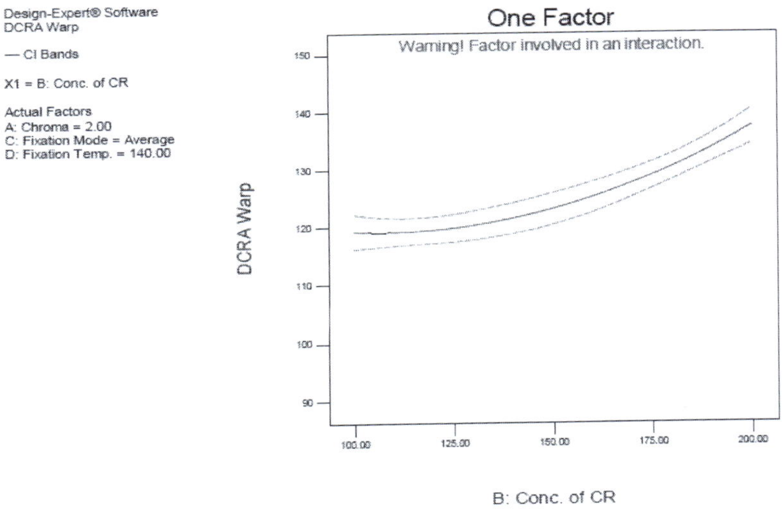

Figure 5. Effect of B on DCRA warp

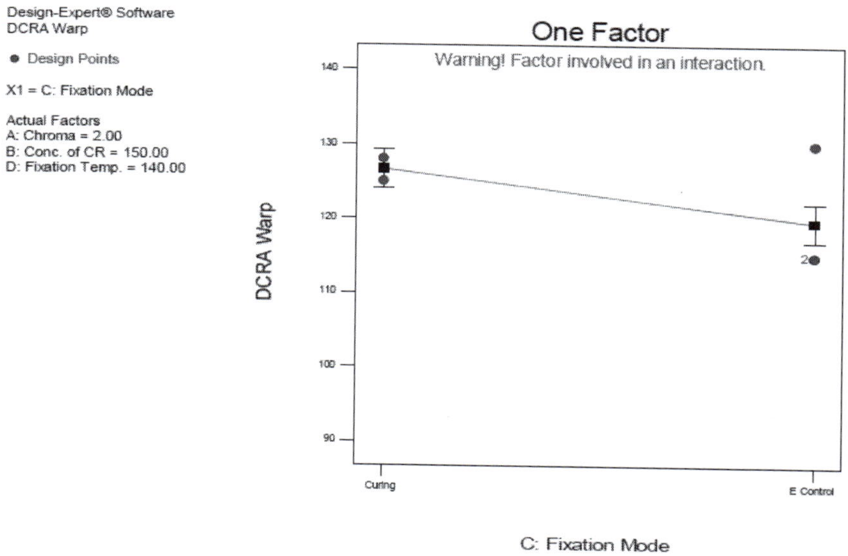

Figure 6. Effect of C on DCRA warp

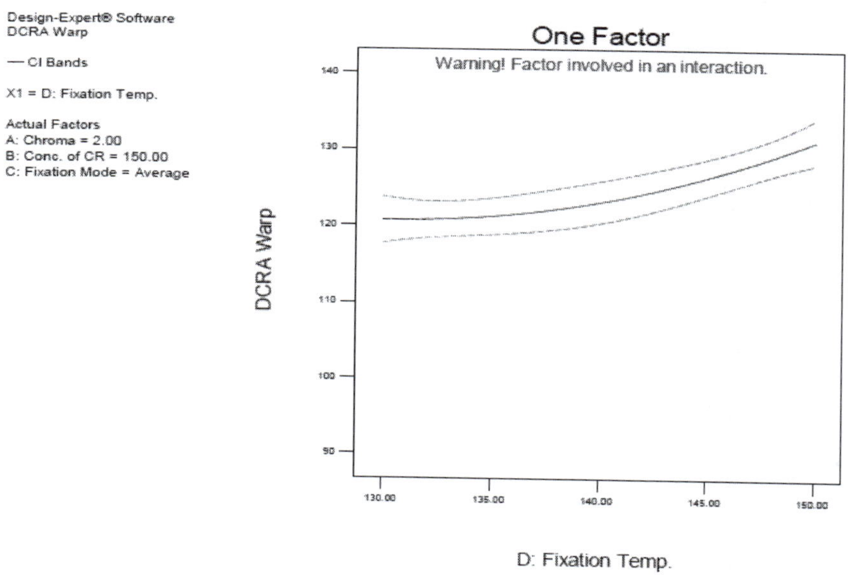

Figure 7. Effect of D on DCRA warp

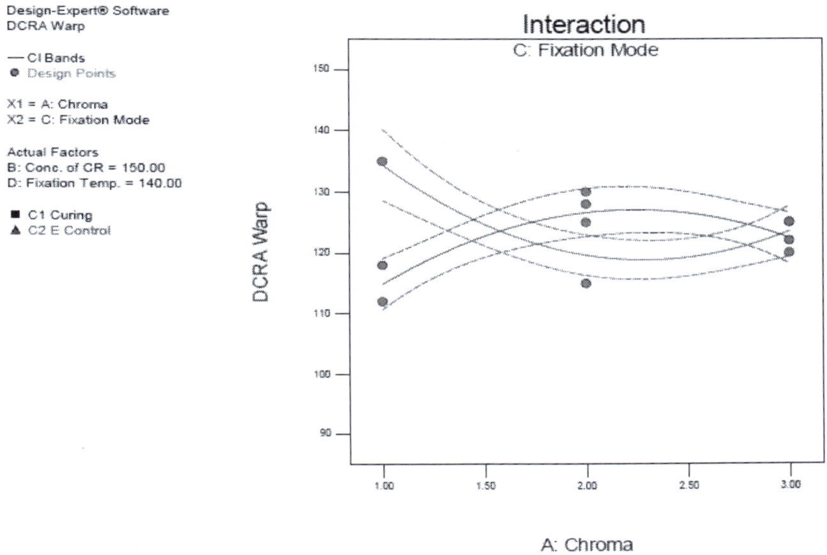

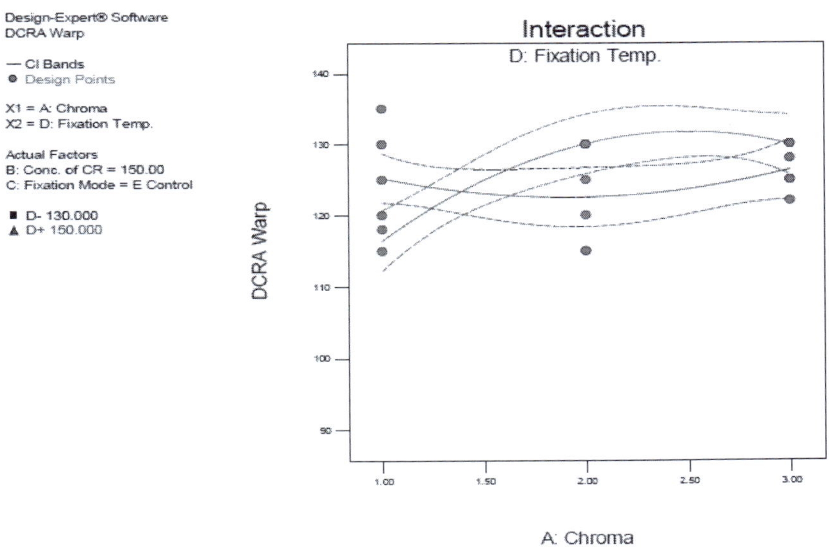

Figure 9. Interaction plot of AD with respect to DCRA warp

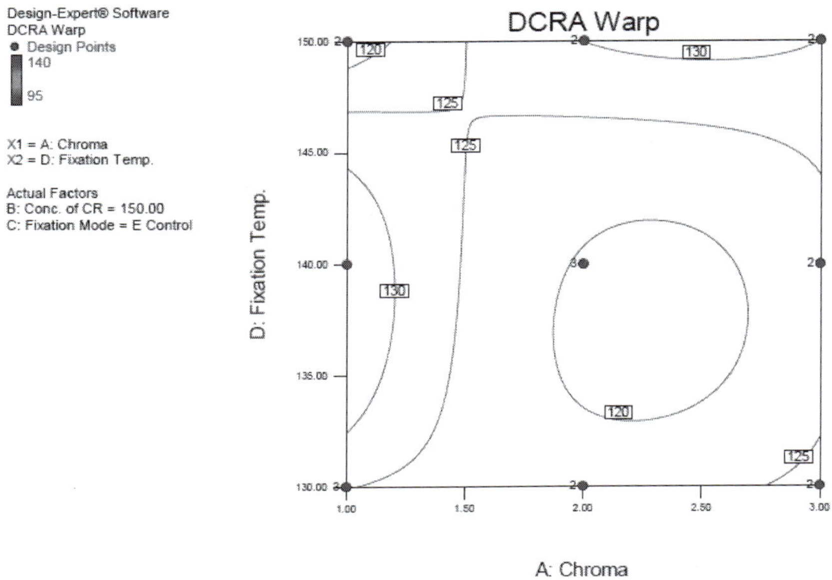

Design-Expert® Software
DCRA Warp
Design Points
140
95
X1 = A: Chroma
X2 = D: Fixation Temp.
Actual Factors
B: Conc. of CR = 150.00
C: Fixation Mode = E Control
DCRA Warp
D: Fixation Temp.
A: Chroma

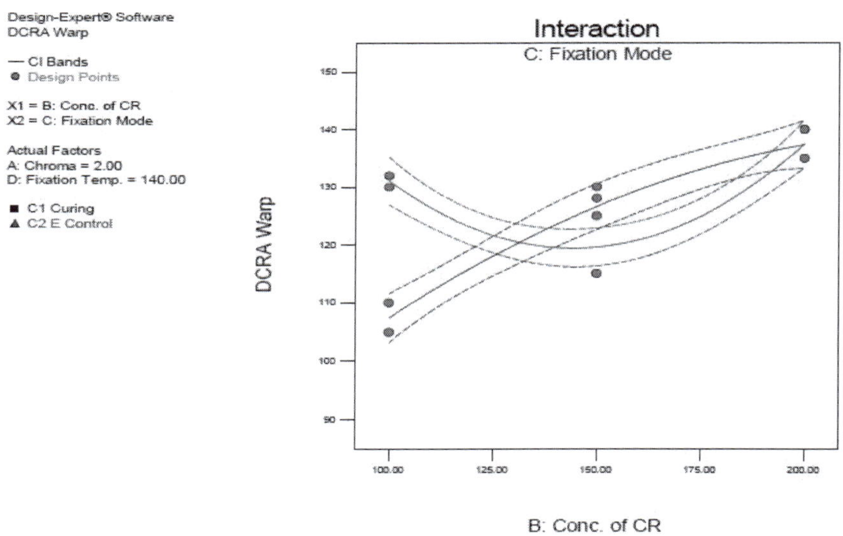

Design-Expert® Software
DCRA Warp
CI Bands
Design Points
X1 = B: Conc. of CR
X2 = C: Fixation Mode
Actual Factors
A: Chroma = 2.00
D: Fixation Temp. = 140.00
C1 Curing
C2 E Control
Interaction
C: Fixation Mode
DCRA Warp
B: Conc. of CR

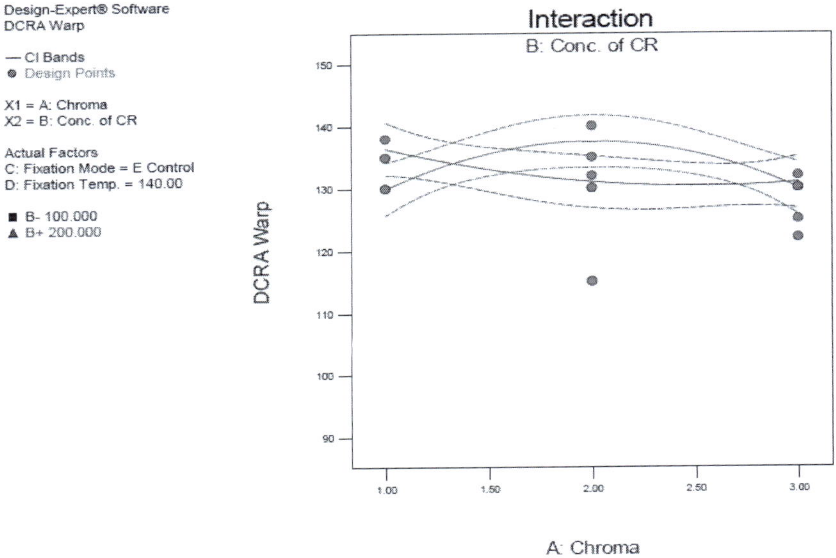

Design-Expert® Software
DCRA Warp
— CI Bands
Design Points
X1 = A: Chroma
X2 = B: Conc. of CR
Actual Factors
C: Fixation Mode = E Control
D: Fixation Temp. = 140.00
B- 100.000
B+ 200.000
Interaction
B: Conc. of CR
DCRA Warp
A: Chroma

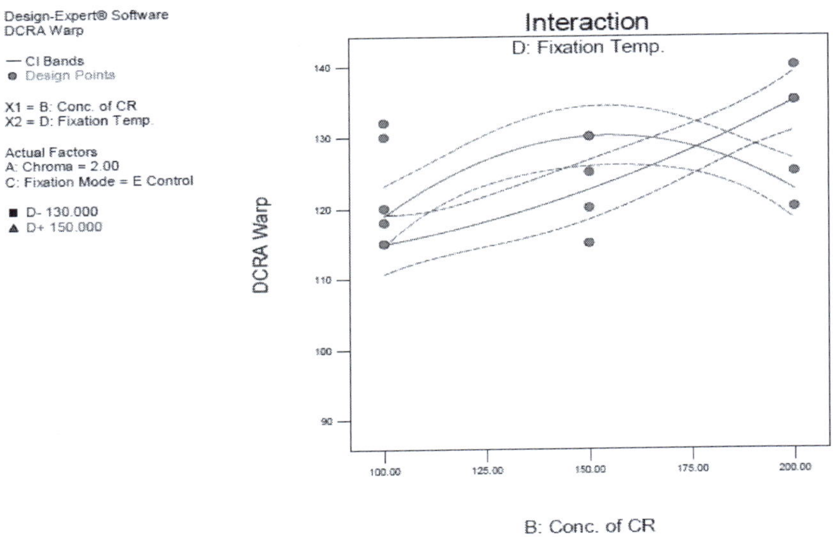

Design-Expert® Software
DCRA Warp
— CI Bands
Design Points
X1 = B: Conc. of CR
X2 = D: Fixation Temp.
Actual Factors
A: Chroma = 2.00
C: Fixation Mode = E Control
D- 130.000
D+ 150.000
Interaction
D: Fixation Temp.
DCRA Warp
B: Conc. of CR

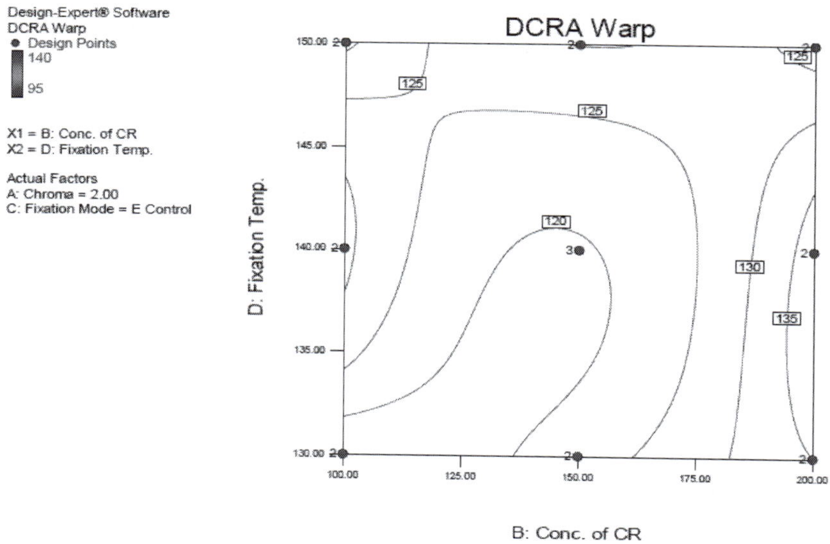

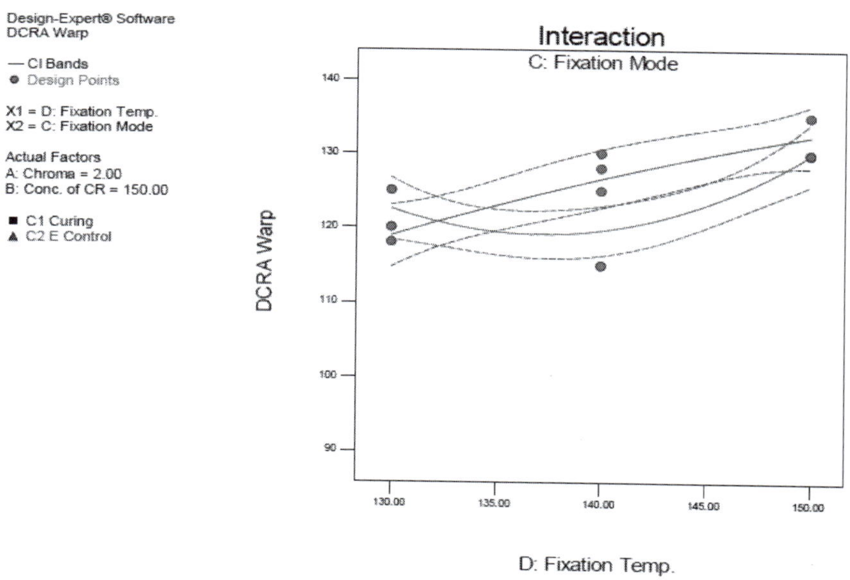

Figure 15. Interaction plot of CD with respect to DCRA warp

4. CONCLUSIONS

The factorial design of experiments for single step fixation process of reactive printing and crease resistance finishing was studied. The quadratic effect of three factors namely, concentration of crease resistant, fixation method and fixation temperature on dry crease recovery angle were identified. The statistical analysis gave a reasonably good fit with an F-value of 24.10 and an R2-value of 0.9587. According to the significance effect obtained in variance analysis, the concentration of crease resistant was the most significant factor for the dry crease recovery angle under the conditions investigated.The information obtained from the mathematical model has been used to create optimal parameters of single step fixation process of reactive printing and crease resistance finishing process.

ACKNOWLEDGEMENTS

The authors wish to express their gratitude to Gul Ahmed Textile Mills Ltd and Clariant Pakistan Ltd, for carrying out the necessary experimental work.

REFERENCES

1. H. Tavanai1, S. M.Taheri, M. Nasiri, "Modelling of Colour Yield in Polyethylene Terephthalate Dyeing with Statistical and Fuzzy Regression", Iranian Polymer Journal, 14 (11), 954-967 (2005)
2. H. Tavanai, A. Z. Hamadani, M. Askari, "Modelling of Colour Yield for Selected Reactive Dyes in Dyeing Cotton Cloth by Two Phase Pad-steam Method", Iranian Polymer Journal, 15 (3), 207-217(2006)
3. V. Vassileva, E. Valcheva, Z. Zheleva, "The Kinetic Model of Reactive Dye Fixation on Cotton Fibres", Journal of the

University of Chemical Technology and Metallurgy, 43(3), 323-326 (2008)

4. F. Asim, M. Mahmood, M. A. Siddiqui, "Optimization of process parameters for simultaneous fixation of reactive printing and crease resistant finishing using desirability function", International Conference on Systems Engineering and Engineering Management, 1098-1105 (2011)

5. F. Asim, M. Mahmood, "Multi response optimization of simultaneous fixation of reactive printing and crease resistant finishing using desirability function", International Journal of Statistics and Applications, 2(6), 126-131 (2012)

6. F. Asim, M. Mahmood, "Optimization of process parameters for simultaneous fixation of reactive printing and crease resistant finishing", Journal of Textile and apparel Technology Management, 7(3), Spring (2012)

7. Y. H. El Hamaky, S. Tawfeek, D. F. Ibrahim, D. Maamoun, S. Gaber, "Printing Cotton Fabrics with Reactive Dyes of High Reactivity from an Acidic Printing Paste", Coloration Technology, 123(6), 365-373 (2007)

8. N. S. E. Ahmed, Y. A. Yourself, R. M. El-Shishtawy, A. A. Mousa, "Urea/Alkali Free Printing of Cotton with Reactive Dyes", Coloration Technology, 122(6), 324-328 (2006)

9. Y. Yang, V. Naarani, "Effect of Steaming Conditions on Colour and Consistency of Ink-Jet Printed Cotton Using Reactive Dyes", Coloration Technology, 120(3), 127-131 (2004)

10. C. W. M. Yuen, S. K. A. Ku, P. S. R. Choi, C. W. Kan, "Study of the Factors Influencing Colour Yield of an Inkjet-Printed Cotton Fabric", Coloration Technology, 120(6), 320-325 (2004)

11. Y. Can, M. Akaydin, Y. Turhan, E. Ay, "Effect of Wrinkle Resistance Finish on Cotton Fabric Properties", Indian Journal of Fibre & Textile Research, 34, 183-186 (2009)

12. M. S. Hassan, Crease Recovery Properties of Cotton Fabrics Modified by Urea Resins under the Effect of Gamma

Irradiation", Radiation Physics and Chemistry, 78(5), 333-337 (2009)

13. F. Asim, M. Mahmood, "Reactive Printing & Crease Resistance Finishing of Cotton Fabrics. Part-I Study of Influential Factors by an Experimental Design Approach", Journal of Textile & Apparel, Technology and Management, 7(1), (2011)

14. F. Asim, M. Mahmood, "Reactive Printing and Crease Resistance Finishing of Cotton Fabrics-Effects of Fixation Modes by 22.41 Mixed Factorial Design", International Journal of Textile Science, 1(6), 94-100 (2012)

15. F. Asim, M. Mahmood, "Effects of Process Parameters on Single Step Fixation of Reactive Printing and Crease Resistance Finishing of Cotton Fabrics using 23 Factorial Design", International Journal of Textile Science, 2(1), February (2013)

16. S.M. Mortazavi, M.A. Alsharif, M. Jannesari, "A study on Simultaneous Fixation of Reactive Dye Printing and Crease Resist Finishing on Cotton Fabric (Effect of DMDHEU Concentration, Types of Catalyst and Drying Conditions)", http://140.134.132.124:8080/dspace/bitstream/2377/3912/1/ce05atc902007000039.pdf

14
CHAPTER

Investigation of Different Washing Effects on Physical and Mechanical Properties of Cotton Knitted Garments

Solaiman[1], Abdur Rouf[2], Shamsuzzaman Rasel[1], Elias Khalil[1]

[1]*Department of Textile Engineering, World University of Bangladesh, Dhaka, Bangladesh*

[2]*Department of Fashion Design and Technology, Shanto-Mariam University of Creative Technology, Dhaka, Bangladesh*

ABSTRACT

This paper deals with three types of washing (enzyme, softener, Silicone) effects on 100% cotton Single Jersey T-shirt, Slub Single Jersey T-shirt, Double Lacoste (5% Lycra) Polo shirt, Single jersey CVC (T-shirt) and PC single Jersey (T-shirt). Typical industrial washing procedures and techniques were followed and then physico-mechanical properties were analyzed under standard condition. It is observed that after washing, weight of each garments has increased from 5% to 36%; stitch length of the constituted fabric has decreased from 1% to 10% approximately; CPI has increased 23% for CVC fabric only for enzyme treatment and for other process it has a little bit or no wash effects. Same effects found for WPI except 100% cotton and Double Lacoste. Shrinkage property both for lengthwise and widthwise remains the same except CVC and Double Lacoste knitted garments and spirality has almost unchanged except 100% cotton and slub fabric. Both the dry and wet rubbing properties have improved after any types of wash except 100% cotton on Silicone wash. Water absorbency property has increased; pH of all the samples is under controlled and lies between 4.5 and 5.0. But there is no significant change of colorfastness to wash and stain after these washing treatments.

Keywords: Washing Effects, Knitted Garments, Physical and Mechanical Properties, pH, Color Fastness

1. INTRODUCTION

Industrial garments washing is one of the major processes followed in the textile sector. By industrial garments washing, dust, dirt and infectious materials can be removed from garments. For improving special look on garments as per fashion requirement, a variety of wash techniques can be followed [1] . For washing of denim garments, a range of treatment methods such as enzymatic treatment [2] , bleaching treatment [3] [4] , acid treatment [5] [6] , and Silicone treatment [7] , are used widely. They all are aimed at new possible effects of fabric appearance. Particularly the dry finishing creates many effects on denim fabric; it will stimulate the customers to buy, and also it increases the market potential of the denim market [8] . Nowadays along with denim washing knit garments such as T-shirt, Polo shirt, and trouser, are washed by using different techniques as follows: enzymes wash, softener wash, Silicone wash, tie dye wash, pigment wash, caustic wash, etc. are used to create or enhance the physical and mechanical property changes [9] [10] . Enzymes are important in regulating the chemical reactions that occur within all living organisms. Without enzymes, many of these reactions would not be able to take place at a perceptible rate [11] . Fabric softener (also called fabric conditioner) is a chemical compound that prevents static cling and confers many other desirable properties to laundry that has been machine-washed. They are available as solutions and solids, sometimes impregnated in dryer sheets [12] . Silicones are polymers that include Silicone together with carbon, hydrogen, oxygen, and sometimes other elements. Some common forms include Silicone oil, Silicones grease, Silicone rubber, Silicones resin, and Silicone caulk [13] . Silicone is the second most abundant element on earth, after oxygen, making up approximately 25 percent of the earth's crust by weight; largely found in the form of oxides such as sand and clay. Sodium silicate, of formula Na_2SiO_3, is commercially produced by the reaction of soda ash and sand in a furnace [14] .

This paper investigates the impact of enzymes, softener and Silicone wash on the physical and mechanical properties of knit garments as these properties determine the wearers feel and life of the end product. The paper also investigates the optimum use of enzyme with the fixed proportion of pumice stones.

2. METHODS AND MATERIALS

2.1. Materials

2.1.1. Garments Samples

Following knits garments are used for the research purpose:

Single Jersey T-Shirt (100% cotton) GSM 140, Single Jersey T-Shirt (Slub yarn) GSM 170, Double Lacoste Polo Shirt (5% Lycra) GSM 228, Single Jersey T-Shirt (CVC Fabric) GSM 180, Single Jersey T-Shirt (PC Fabric) GSM 138.

2.1.2. Chemicals

Enzymes (RUCOLASE ZSS), Silicone (DYCHUFIX BLACK B, double components), Softener (Cationic) (Softycell IPS).

2.1.3. Washing Machine

AZIZ Metal Engineering Works Ltd, M/C capacity-60 kg, Max speed-50 k.m., temp 100°C, Heating source- Steam.

2.2. Methods

2.2.1. Enzyme Wash

At first all the five types of the knit fabrics samples are taken to the washing machine. Here material and liquor ratio is 1:5. Approximately 0.08% enzymes (RUCOLASE ZSS), 0.2% acetic acid and 0.06% detergent (WBL) is used. The washing process carried out for 15 minutes at a temperature of 45°C. Then rinse the samples for 2 minutes. Finally tumble dryer is used to dry the fabric samples.

2.2.2. Garments Wash (with Softener)

At first all the five types of the knit fabrics samples are taken to the washing machine. Here material and liquor ratio is 1:6. Approximately 0.08% softener (Softycell IPS) is used. The washing process carried out for 10 minutes at a temperature of 60°C. Finally tumble dryer is used to dry the fabric samples.

2.2.3. Silicone Wash

At first all the five types of the knit fabrics samples are taken to the washing machine. Here material and liquor ratio is 1:8. Approximately 0.1% Silicone (DYCHUFIX BLACK B) is used during the process. The washing process carried out for 10 minutes at a temperature of 60°C. Finally tumble dryer is used to dry the fabric samples.

2.2.4. Determination of Fabric Weight

Fabric weight was measured according to ASTM D 3776 methods of standards. We cut three pieces of the fabric with the GSM cutter (gram per square inch) and Weight the fabric with the electric balance. Then multiplied by 100 and took the average results [15] .

2.2.5. Determination of Stitch Length

According to BS EN 14970-2006 standard we measured the stitch length. At first we count the length of the fabric in centimeter of different number of needle. Here for different types of fabric needs different number of needle count for different knit fabric. Needs 100 needles count for every type of single jersey fabrics. Needs 100 needles count for all types of Rib fabrics. Therefore Interlock need 50 needles count for stitch length measurement [16] .

2.2.6. Determination of Course per Inch (CPI)

CPI of the fabric is calculated by counting the number of the yarn (Course) contents in 1 inch of the fabric. For this the fabric is marked by 1 inch in rectangular area. Then the yarn is open out and count the number courses contains in 1 inch [17] .

2.2.7. Determination of Wales per Inch (WPI)

WPI of the fabric is calculated by counting the number of the yarn (Wales) contents in 1 inch of the fabric. For this the fabric is marked by 1 inch in rectangular area. Then the yarn is open out and count the number Wales contains in 1 inch [17] .

2.2.8. Determination of Shrinkage

According to ASTM D 2259 shrinkage of these sample garments were tested. At first we Cut 3 specimens, each 50 × 50 cm^2. Then marked the fabric 35 cm × 35 cm, inside 50 cm × 50 cm are of fabric. Then the fabric samples are boiled for 10 - 15 minutes with distilled water. After boiling tumble dryer is used to dry that fabric. Finally, length and width of the fabric sample is measured [18] .

Shrinkage% = (before wash ? after wash)/before wash.

2.2.9. Determination of Dimensional Stability

According ISO 3759 standards fabric samples were tested. The fabric samples were cut at a square of 50 × 50cm. we laid the fabric to relax for 4 hours on a bench in ambient conditions so that it became smooth and tension free. We placed the template on the fabrics samples with the side of the template parallel to the length (warp) direction. The mark the three width and length marks in pairs 35cm apart. Then draw around the edge of the template, do not round off the corners. Then we draw an arrow outside the measurement area to denote the length (warp) direction prior to cutting from the main piece. By using ECE detergent and Sodium Perborate at a temperature of 60°C a continuous process of washing techniques are carried out for 10 - 15 minutes. Finally we dried the fabric samples and measured the changes of dimensional stability by comparing with scale [19] .

2.2.10. Determination of Rub Property

According to AATCC 8 standards both the dry and wet rub test were done as follows:

1) Dry rubbing test:

A test sample is clamped to the instrument base and a square of standard crocking cloth is fixed to a 16 mm diameter, acrylic rubbing finger. The finger rests on the sample with a pressure of 900 grams force and traverses a straight path approximately 100 mm long with each stroke of the arm. After testing textile materials, the crocking cloth is removed and compared against the AATCC Gray Scale for Staining or Chromatic Transference Scale. Based on this comparison, we assign a grade to the color change or amount of color transfer.

2) Wet rubbing test:

Repeat the procedure from dry rubbing test on another sample with the white cotton test cloth wetted it in distilled water. Ensure that the rubbing test cloth will be wetted with water to 100% take up.

Compare the contrast between the untreated and treated white rubbing cloth with the staining. Grey scale rates from 1 - 5 and changing in the tested specimen with changing grey scale. This visual assessment is done in a color matching cabinet under standard lighting D-65 [20] .

2.2.11. Determination of Absorbency

For absorbency testing we followed AATCC 79 methods of standards. Samples were placed in an embroidery hoop with all creases out of it. A burette dispenses a drop of water onto the surface of the fabric from a distance of 9.5 mm below the burette. Time was recorded until the water drop absorbs completely [21] .

2.2.12. Determination of pH (AATCC 81-2006)

We conducted this test according to the AATCC 81 methods. We used 10 g specimen, cut into small pieces. Then Boiled 250 ml distilled water for 10 minutes, immerse the specimen, cover the beaker with a watch glass and boiled for an additional 10 minutes. Allow the covered beaker and contents to cool down to room temperature. Finally we determined the pH of the extract using a pH meter [22] .

2.2.13. Determination of Colorfastness to Washing Test (BS EN ISO 105 C06)

The samples were tested according to BS EN ISO 105 C06 and in the delivered condition; didn't wash and/or tumble it before testing. We used only multifibre from James Heal. The washing solution prepared freshly (one litre). The current liquor ratio (liquor/sample) is 20:1. We put the grey frame in same grey colour as background. White paper were placed as background or according to the test method one or more original multifibre to assure no transparent effect coming out of the background. The edges of the specimen and original multifibre were cut straight. The edges of the specimen and original multifibre had cut straight under defined light D65 with defined angle of 45° [23] .

3. RESULTS AND DISCUSSIONS

3.1. Effects of Fabric Weight after Washing on Different Knit Garments

A slight increase of the GSM of all the 5 types of knit garments are found from the experiment of the different 3 types of the wash experiments. A significant amount of the GSM varies when the knit garments are applied with enzyme wash techniques though all the other washing process also affects the changes of the GSM of the knit garments. Exceptionally only GSM of the T-shirt which is composed of the CVC fibre decreases. Table 1 has shown this.

3.2. Effects on Stitch Length after Washing of Different Knit Garments

The following Table 2 illustrates the complete decrease of the S.L. after washing of the knit fabrics. Approximately all types of wash have the same wash effects over all types of knit garments. However, S/J 100% cotton knit garments has the dominant changes than others.

Table 1. Effects on fabric weight after washing of different knit garments.

Parameters	S/J 100% Cotton Knit Garments T-Shirt	S/J Slub Knit Garments T-Shirt	Double Lacoste 5% Lycra Knit Garments Polo Shirt	S/J CVC Knit Garments T-Shirt	S/J PC Knit Garments T-Shirt
Before Wash	140	170	228	180	138
After Enzyme Wash	190	198	250	189	160
After Silicone Wash	185	185	238	165	154
After Softener Wash	177	190	244	192	160

3.3. Effects on CPI after Washing of Different Knit Garments

CPI has a great number of increment found in enzyme wash compared to all others washing process. Interestingly 23% of the CPI increases for CVC knit fabric garments and 17% for Double Lacoste knit fabric garments after enzyme wash. All others have a very small increment. A littile bit increment found in Silicone wash but garments wash has a very less or no increment for CPI fromFigure 1.

3.4. Effects on WPI after Washing of Different Knit Garments

All the wash has a great amount of WPI changing effects (Figure 2). This means that these are the responsible for the increment of the fabric GSM after washing. Here CVC fabric garments has a higher amount of WPI changing effects comparing to others, second one is S/J 100% cotton knit garments.

Table 2. Effects on stitch length after washing of different knit garments.

Parameters	S/J 100% Cotton Knit Garments T-shirt	S/J Slub Knit Garments T-Shirt	Double Lacoste 5% Lycra Knit Garments Polo Shirt	S/J CVC Knit Garments T-Shirt	S/J PC Knit Garments T-Shirt
Before Wash	29.5 cm	28.5 cm	33.0 cm	25.5 cm	27.5 cm
After Enzyme Wash	26.8 cm	28.0 cm	31.0 cm	25.0 cm	25.8 cm
After Silicone Wash	26.6 cm	27.5 cm	32.0 cm	28.0 cm	26.8 cm
After Softener Wash	27.0 cm	27.8 cm	31.5 cm	25.2 cm	26.6 cm

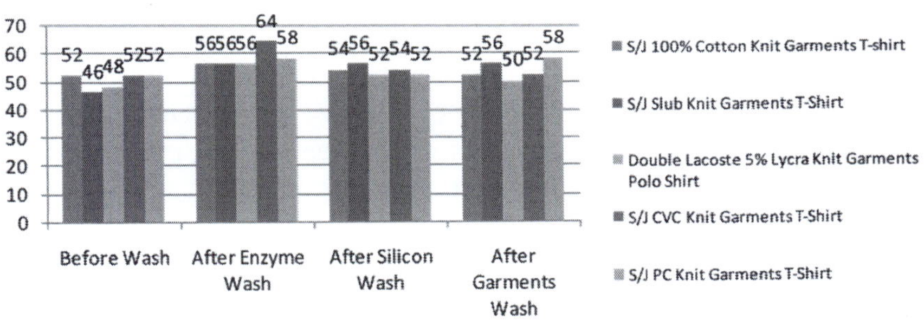

Figure 1. Effects on CPI after washing of different knit garments.

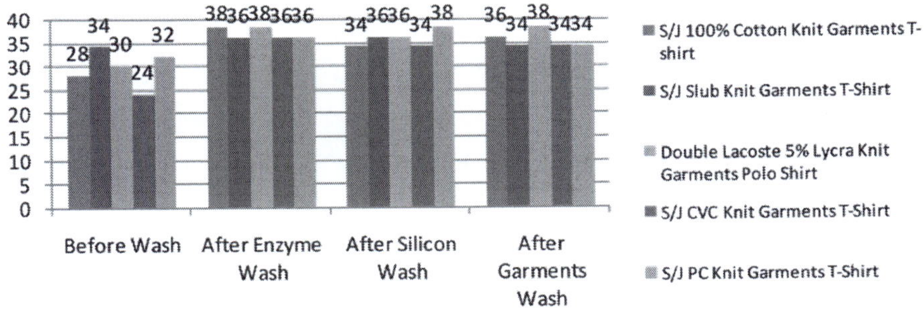

Figure 2. Effects on WPI after washing of different knit garments.

3.5. Effects of Shrinkages after Washing on Different Knit Garments

A very little or no changes are found both for lengthwise and widthwise shrinkage (Table 3 andTable 4). The best quality shrinkage found for S/J Slub and PC fabric and found only 3% of the widthwise shrinkage, whereas others shrink quality also improved. Silicone wash has a little change of shrinkage, only 3% for the fabrics but S/J 100% cotton, D. Lacoste and PC fabric though the shrink quality compared to before wash decreases. Garments wash has a little changes of shrinkage, only 3% for the fabrics but S/J 100% cotton, D. Lacoste and PC fabric though the shrink quality compared to before wash decreases.

3.6. Effects on Dimensional Stability after Washing of Different Knit Garments

After enzyme wash spirality of all the fabrics increases (Table 5) which decrease the quality of knit garments. Similarly dimensional stability also increases that is the quality for the fabrics decreases after silicon wash. Also garments wash has an increment of the spirality of the fabrics but the fabrics quality remains same as it was before.

3.7. Effects on Rub Property after Washing of Different Knit Garments

Most of the cases rub property (Table 6) of the fabrics improved. Much improvement found for CVC fabrics. Others remains same or improved a little bit by measuring the rub property through comparing the measuring the scale.

Only a little bit changes are found when all the fabrics are under wet rub (Table 7). Though no severe changes are found but the quality of the fabrics improved.

Table 3. Effects on lengthwise shrinkage after washing on different knit garments.

Parameters	S/J 100% Cotton Knit Garments T-Shirt	S/J Slub Knit Garments T-Shirt	Double Lacoste 5% Lycra Knit Garments Polo Shirt	S/J CVC Knit Garments T-Shirt	S/J PC Knit Garments T-Shirt
Before Wash	5%	4%	5%	6%	3%
After Enzyme Wash	5%	5%	3%	5%	4%
After Silicone Wash	5%	5%	3%	5%	4%
After Softener Wash	4%	5%	4%	4.5%	2%

Table 4. Effects on widthwise shrinkage after washing of different knit garments.

Parameters	S/J 100% Cotton Knit Garments T-Shirt	S/J Slub Knit Garments T-Shirt	Double Lacoste 5% Lycra Knit Garments Polo Shirt	S/J CVC Knit Garments T-Shirt	S/J PC Knit Garments T-Shirt
Before Wash	4%	2%	3%	4%	2%
After Enzyme Wash	4%	3%	4%	4%	3%
After Silicone Wash	4%	3%	4%	4%	3%
After Softener Wash	4%	3%	3%	4%	3%

Table 5. Effects on dimensional stability after washing on different knit garments.

Parameters	S/J 100% Cotton Knit Garments T-Shirt	S/J Slub Knit Garments T-Shirt	Double Lacoste 5% Lycra Knit Garments Polo Shirt	S/J CVC Knit Garments T-Shirt	S/J PC Knit Garments T-Shirt
Before Wash	4%	3%	3%	4%	2%
After Enzyme Wash	5%	5%	3%	3%	3%
After Silicone Wash	6%	5%	4%	4%	3%
After Softener Wash	5%	4%	3%	4%	3%

Table 6. Effects on dry rub property after washing of different knit garments.

Parameters	S/J 100% Cotton Knit Garments T-Shirt	S/J Slub Knit Garments T-Shirt	Double Lacoste 5% Lycra Knit Garments Polo Shirt	S/J CVC Knit Garments T-Shirt	S/J PC Knit Garments T-Shirt
Before Wash	4.5	4.5	4.0	4.0	4.5
After Enzyme Wash	5.0	4.5	4.5	4.5	4.5
After Silicone Wash	4.5	5.0	4.5	4.5	5.0
After Softener Wash	5.0	5.0	4.5	4.5	5.0

Table 7. Effects on wet rub property after washing of different knit garments.

Parameters	S/J 100% Cotton Knit Garments T-Shirt	S/J Slub Knit Garments T-Shirt	Double Lacoste 5% Lycra Knit Garments Polo Shirt	S/J CVC Knit Garments T-Shirt	S/J PC Knit Garments T-Shirt
Before Wash	4.5	4.0	3.5	3.5	4.0
After Enzyme Wash	4.5	5.0	4.0	4.5	4.5
After Silicone Wash	4.0	4.5	4.0	4.0	4.0
After Softener Wash	5.0	4.5	4.5	4.0	5.0

3.8. Effects on Absorbency after Washing of Different Knit Garments

Absorbency of the fabrics varies a lot (Figure 3). For S/J 100% cotton and S/J Slub fabrics absorbency decreases continuously after all other washing techniques whereas PC fabrics has a fluctuated changes of absorbency on all these 3 washing process.

3.9. Effects on pH Value after Washing of Different Knit Garments

pH has great number of changes (Table 8) than before wash for all types of wash. For enzyme wash pH decreases for all the fabrics but only PC fabrics has little bit increases of the pH. Garments wash has a little bit decreases of the pH for first three types of the fabrics but CVC and PC has remained the same number of pH. Similar tendency found for every types of the fabrics when treated with Silicone wash.

3.10. Effects on Color Changes after Washing of Different Knit Garments

Color change after wash has no or a little bit changes found after washing of the fabrics (Table 9). For 100% cotton S/J fabric has a slight improvement found when treated with Silicone but for other wash no improvement found compared to others. For Slub S/J a little color changes after all types of wash. CVC and PC have the same tendency as shown by S/J Slub fabrics. Only Double Lacoste has a slight improvement for any kinds of wash effects.

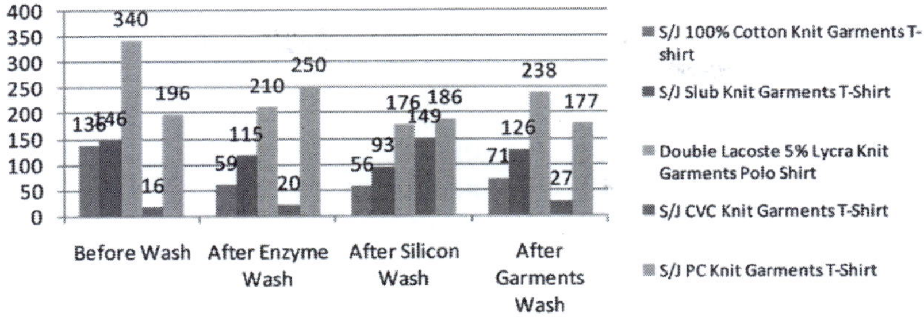

Figure 3. Effects on absorbency after washing of different knit garments.

Table 8. Effects on pH value after washing of different knit garments.

Parameters	S/J 100% Cotton Knit Garments T-Shirt	S/J Slub Knit Garments T-Shirt	Double Lacoste 5% Lycra Knit Garments Polo Shirt	S/J CVC Knit Garments T-Shirt	S/J PC Knit Garments T-Shirt
Before Wash	5.5	5.5	5.5	5.0	4.5
After Enzyme Wash	5.0	4.5	4.5	4.5	5.0
After Silicone Wash	5.0	4.5	5.0	5.0	4.5
After Softener Wash	4.5	4.5	4.5	5.0	4.5

Table 9. Effects on color changes after washing of different knit garments.

Parameters	S/J 100% Cotton Knit Garments T-Shirt	S/J Slub Knit Garments T-Shirt	Double Lacoste 5% Lycra Knit Garments Polo Shirt	S/J CVC Knit Garments T-Shirt	S/J PC Knit Garments T-Shirt
Before Wash	4.5	5.0	4.0	4.5	4.5
After Enzyme Wash	4.5	4.5	4.5	4.5	4.5
After Silicone Wash	5.0	4.5	5.0	5.0	4.5
After Softener Wash	4.5	4.5	4.5	4.5	4.5

Table 10. Effects on color staining after washing of different knit garments.

Parameters	S/J 100% Cotton Knit Garments T-Shirt	S/J Slub Knit Garments T-Shirt	Double Lacoste 5% Lycra Knit Garments Polo Shirt	S/J CVC Knit Garments T-Shirt	S/J PC Knit Garments T-Shirt
Before Wash	4.5	4.5	4.5	4.5	4.5
After Enzyme Wash	4.5	4.5	4.0	4.5	4.0
After Silicone Wash	4.5	4.5	5.0	4.5	4.5
After Softener Wash	4.5	4.5	4.5	4.5	4.5

3.11. Effects on Color Staining after Washing of Different Knit Garments

There are no big changes founds of the color staining after wash experiment. 100% cotton, CVC, D. Lacoste and Slub fabrics has no or a very slight changes of the color staining (from Table 10).

4. CONCLUSION

It has been found that when enzyme wash is applied on the knit garments, it improves all the tested properties of knit garments, and it also reduces the hairy fibers from the fabric surface. Similar results are found in case of Silicone and softener wash as well. Most important results are found for 100% cotton knit garments. Overall all the knit garments become improved after washing. Again with the increasingly important requirement for textile manufacturers to reduce pollution in textile production, the use of enzymes, softeners and Silicones in the chemical processing of fibers and textiles is rapidly gaining wider recognition because of their non-toxic and eco-friendly characteristics. It can be used safely in a wide selection of textile processes. Nowadays application of different types of washing is increasing day by day for preparing a well good enough knit garments.

REFERENCES

1. Mondal, M.I.H. and Khan, M.M.R. (2014) Characterization and Process Optimization of Indigo Dyed Cotton Denim Garments by Enzymatic Wash. Fashion and Textiles, 1, 1-12.
2. Sarkar, J., Khalil, E. and Solaiman, M. (2014) Effect of Enzyme Washing Combined with Pumice Stone on the Physical, Mechanical and Color Properties of Denim Garments.

International Journal of Research in Advent Technology, 2, 65-68.

3. Sarkar, J. and Khalil, E. (2014) Effect of Industrial Bleach Wash and Softening on the Physical, Mechanical and Color Properties of Denim Garments. IOSR Journal of Polymer and Textile Engineering, 1, 46-49.

4. Khan, M.M.R., Mondal, M.I.H. and Uddin, M.Z. (2011) Effect of Bleach Wash on the Physical and Mechanical Properties of Denim Garments. International Conference on Mechanical Engineering, 3.

5. Haq, U.N. and Khan, M.M.R. (2014) Technology of Acid Wash on Woven Denim Apparel with Damp Pumice Stone.

6. Khalil, E. (2015) Effect of Processing Time and Concentration of Potassium Permanganate on Physico-Mechanical Properties of Denim Jeans during Stone Washing. Science Innovation, 3, 68-71.

7. Kashem, P.M. (2008) Garments Merchandising. Luckey-One Traders, Bangladesh.

8. Khalil, E. and Islam, M.M. (2015) Wrinkle Finish on Denim by Resin Treatment: A Review. AASCIT Communication, 2, 82-87.

9. Khalil, E., Sarkar, J., Rahman, M. and Solaiman, M. (2014) Influence of Enzyme and Silicone Wash on the Physico-Mechanical Properties of Non-Denim Twill Garments. International Journal of Scientific & Technology Research, 3, 231-233.

10. Gokarneshan, N., Durairaj, C., Krishnamurthy, P., Shanmugasundaram, S., Subhash, R. and Su, P. (2009) Chemical Finishing and Washing of Knit Wear. http://www.fibre2fashion.com/industry-article/23/2210/chemical- finishing-and-washing-of-knit-wear1.asp

11. http://abyss.uoregon.edu/~js/glossary/enzyme.html

12. Fabric Softener (2015) In Wikipedia, the Free Encyclopedia. https://en.wikipedia.org/w/index.php?title=Fabric_softener&oldid=687529481

13. Silicone (2015) In Wikipedia, the Free Encyclopedia. https://en.wikipedia.org/w/index.php?title=Silicone&oldid=680713579

14. David, A.E. (2004) Immobilization of Enzymes on Nanoporous, Silica Composites.

15. ASTM D3776/D3776M-09a (2013) Standard Test Methods for Mass per Unit Area (Weight) of Fabric. ASTM International, West Conshohocken.

16. BS EN 14970:2006 (2006) Textiles. Knitted Fabrics. Determination of Stitch Length and Yarn Linear Density in Weft Knitted Fabrics.

17. IS 1963 (1982) Methods for Determination of Threads per Unit Length in Knitted Fabrics. Bureau of Indian Standards, New Delhi.

18. AATCC Test Method 96 (2009) Dimensional Changes in Commercial Laundering of Woven and Knitted Fabrics except Wool. American Association of Textile Chemists and Colorists, Research Triangle Park.

19. ISO 3759:2011 (2011) Textiles—Preparation, Marking and Measuring of Fabric Specimens and Garments in Tests for Determination of Dimensional Change.

20. ISO 105-X12:2001 (2001) Textiles—Tests for Color Fastness—Part X12: Color Fastness to Rubbing.

21. AATCC Test Method 79 (2009) Absorbency of Textiles. American Association of Textile Chemists and Colorists, AATCC Technical Manual, Research Tri U.S.A.

22. AATCC Test Method 81-2006 (2006) pH of the Water Wet Processed Textiles. American Association of Textile Chemists and Colorists, AATCC Technical, Research Triangle Park, North Carolina.

23. ISO 105-C10:2006 (2006) Textiles—Tests for Colour Fastness—Part C10: Colour Fastness to Washing with Soap or Soap and Soda.

Index